Growing and Handling of Bacterial Cultures

Edited by Madhusmita Mishra

Published in London, United Kingdom

IntechOpen

Supporting open minds since 2005

Growing and Handling of Bacterial Cultures
http://dx.doi.org/10.5772/intechopen.74273
Edited by Madhusmita Mishra

Contributors
Branka Bedenić, Sanda Sardelić, Galina Novik, Olga Meerovskaya, Victoria Savich, Julia Oxford, Lisa Warner, Olga Mass, Nancy Lenn, Briana Grantham, Ralf Pörtner, Rebecca Faschian, Yasir Mehmood, Manal Mohammed, Beata Orzechowska, Angel Manteca, Beatriz Rioseras, Nathaly González-Quiñónez, Paula Yagüe, Gemma Fernández-García

First published in London, United Kingdom, 2019 by IntechOpen
IntechOpen is the global imprint of INTECHOPEN LIMITED, registered in England and Wales, registration number: 11086078, 7th floor, 10 Lower Thames Street, London, EC3R 6AF, United Kingdom
Printed in Croatia

British Library Cataloguing-in-Publication Data
A catalogue record for this book is available from the British Library

Additional hard and PDF copies can be obtained from orders@intechopen.com

Growing and Handling of Bacterial Cultures
Edited by Madhusmita Mishra
p. cm.
Print ISBN 978-1-78985-846-4
Online ISBN 978-1-83880-123-6
eBook (PDF) ISBN 978-1-83880-124-3

Meet the editor

Dr. Madhusmita Mishra holds a PhD and has more than 7 years of experience in academic and industrial Research and Development. She has extensive experience in Biotechnology, Microbiology, Molecular Biology, Analytical Instrumentation and associated Data Stations, and Data Analysis. Her doctoral work covered both aerobic and anaerobic microbiology along with wide applications of molecular biology tools. As a postdoctoral fellow, she worked in field of development of microdevices for health diagnostics, a project funded by Grand Challenges Canada (GCC). Dr. Mishra has also worked as Analytical Researcher in one of the major oil companies, Shell Technology India Market Private Ltd. Her recent role as experimentalist at Shell involves project designing and implementation of analytical instrumentation for characterization of crude oil and its products.

Contents

Preface

This Edited Volume is a collection of reviewed and relevant research chapters, concerning the developments within the Growing and Handling of Bacterial Cultures. The book includes scholarly contributions by various authors and edited by a group of experts pertinent to Life Sciences. Each contribution comes as a separate chapter complete in itself but directly related to the book's topics and objectives.

The book includes chapters dealing with the topics: Geobacillus Bacteria - Potential Commercial Applications in Industry, Bioremediation, and Bioenergy Production, Growing and Handling of Bacterial Cultures within a Shared Core Facility for Integrated Structural Biology Program, Carbapenemases, Design and Operation of Fixed-Bed Bioreactors for Immobilized Bacterial Culture, What Is Limulus Amebocyte Lysate (LAL) and Its Applicability in Endotoxin Quantification of Pharma Products, The War between Bacteria and Bacteriophages, and Mycelium Differentiation and Development of Streptomyces in Liquid Nonsporulating Cultures: Programmed Cell Death, Differentiation, and Lysis Condition Secondary Metabolite Production.

The target audience comprises scholars and specialists in the field.

IntechOpen

Chapter 1

Geobacillus Bacteria: Potential Commercial Applications in Industry, Bioremediation, and Bioenergy Production

Galina Novik, Victoria Savich and Olga Meerovskaya

Abstract

The genus *Geobacillus* is represented by obligately thermophilic bacteria able to grow in the temperature range of 35–75°C. They are modest bacteria isolated from various sources on routine media such as nutrient agar. Originally classified as representatives of *Bacillus*, the species of *Geobacillus* were established in 2001 as a new genus. However, sequence similarity between all species indicates that at least some species need to be reclassified at the genus level. In addition to 16S rRNA, housekeeping genes, 16S-23S rRNA gene internal transcribed spacer, and repetitive sequences can be used in classification and identification of thermophilic bacteria. The ability to survive and grow at high temperatures as well as utilization and synthesis of a wide range of compounds makes these bacteria and their products attractive for use in various spheres: food, paper, biotechnology industries, medicine, bioremediation, etc. A broad spectrum of applications arouses increased interest in the study of physiological and biochemical characteristics and triggers emergence of new usage areas for *Geobacillus*, such as bioenergy. The growing demand for energy leads to the development of alternative technologic options. *Geobacillus* species demonstrated the ability to generate or enhance productivity of important sources of bioenergy such as ethanol, isobutanol, 2,3-butanediol, biodiesel, and biogas.

Keywords: *Geobacillus* bacteria, biotechnology industry, production of thermostable enzymes, food enzyme applications, bioenergy, biogas, technologies, energy efficiency

1. Introduction

The *Geobacillus* species are Gram-positive, aerobic or facultatively anaerobic, spore-forming, rod-shaped cells with the temperature range for growth 35–75°C (optimum at 55–65°C) (**Figure 1**). Neutrophilic bacteria multiply at pH 6.0–8.5, with optimal pH values 6.2–7.5. Most species are modest bacteria able to develop without growth factors or vitamins and to utilize n-alkanes as carbon and energy sources [1].

As obligate thermophiles, *Geobacillus* might have been expected to be found only in the warmest regions of the planet, such as equatorial deserts or naturally

Figure 1.
Cell morphology of Geobacillus stearothermophilus (basonym Bacillus stearothermophilus) BIM B-202 from Belarusian collection of nonpathogenic microorganisms. Legend: phase-contrast microscopy, magnification 1000×, 16 h cultivation; → – spores.

occurring geothermal and hydrothermal springs. However, *Geobacillus* can be isolated in large numbers anywhere, even from cool soils and permanently cold ocean sediments. It was shown that these bacteria are also minor opportunistic decomposers of plant-derived organic matter, capable of rapid growth under transient thermophilic conditions, but endowed with mechanisms to survive extended time spans when growth is impossible. Nevertheless, *Geobacillus* species play only a modest role in large microbial communities. High populations of bacteria have gradually accumulated in long-term perspective due to spore formation. Adaptive features of *Geobacillus* spores ensure their mobilization in the atmosphere and transport over long distances. Their spores are suggested to remain viable for long periods of time [2]. Spores of *Bacillus* species related to geobacilli have shown resistance to heat, radiation, and chemicals [3].

Growth at high temperatures makes *Geobacillus* species promising agents in biotechnological processes. They can be sources of various thermostable enzymes, such as proteases, amylases, lipases, and pullulanases. *Geobacillus* species can also generate exopolysaccharides and bacteriocins and take part in production of biofuel and bioremediation. New applications are constantly emerging for this group of thermophilic bacteria.

2. Cultivation of *Geobacillus* bacteria

Geobacillus species are obligately thermophilic chemoorganotrophs. Temperature ranges for growth generally lie between 37 and 75°C, with optima between 55 and 65°C. They are neutrophilic bacteria growing within pH range 6.0–8.5 at optimal values 6.2–7.5. Growth factors, vitamins, NaCl, and KCl are not required; So that, most strains will grow on routine media such as nutrient agar. A wide range of substrates is utilized, including carbohydrates, organic acids, peptone, tryptone, and yeast extract. The ability to utilize hydrocarbons as carbon and energy sources is a widely distributed property in the genus [1].

Thermophiles may be obtained easily by incubating environmental or other samples in conventional cultivation media at high temperatures. A selective method for the isolation of thermophilic flat sour organisms from food was described in 1963 by Shapton and Hindes. The method used yeast-glucose-tryptone agar containing peptone (5 g), beef extract (3 g), tryptone (2.5 g), yeast extract (1 g), and glucose (1 g) diluted in distilled water (1000 ml) [4]. This procedure was used for the isolation of *Geobacillus stearothermophilus* [1]. A prototrophic strain of *G. stearothermophilus* has been shown to grow in the medium containing only glucose and mineral salts, while auxotrophic strains additionally required biotin, thiamine, nicotinic acid, and DL-methionine. The presence of L-leucine in minimal medium necessitated the addition of L-valine; however, growth occurred in the absence of both amino acids [5]. The use of medium containing 0.5% beef extract, 0.9% soy peptone, 0.2% NaCl, 0.1% K_2HPO_4, and 0.075% KH_2PO_4 resulted in 10 times higher biomass production by *G. stearothermophilus* than the application of the standard fermentation medium [6].

Geobacillus caldoxylosilyticus was isolated from soil by adding 0.1–0.2 g sample to the minimal medium and incubating at 65°C for up to 24 h. After two transfers of 1 ml culture into the fresh medium, enrichments were plated on solidified minimal medium and incubated at 65°C for 24 h [7]. Further isolations were made by heating samples at 90°C for 10 min, plating on CESP agar (casitone, 15 g; yeast extract, 5 g; soytone, 3 g; peptone, 2 g; $MgSO_4$, 0.015 g; $FeCl_3$, 0.007 g; $MnCl_2 \cdot 4H_2O$, 0.002 g; water, 1000 ml; pH, 7.2.) and incubating at 65°C for 24 h [8]. The strain *G. caldoxylosilyticus* UTM6 demonstrated ability to reduce toxic chromium (VI) to nonharmful chromium (III). It was found to grow optimally in nutrient broth medium supplemented with 250 ppm of glucose at 55°C and pH 6.5, with the highest OD_{600} reading of 0.910 [9].

Geobacillus gargensis was isolated from the upper layer of a microbial mat of Garga hot spring by serial dilutions and inoculation onto the agar medium supplemented with 15 mM sucrose: TES [N-tris(hydroxymethyl)methyl-2-aminoethanesulfonic acid], 10 g; NH_4Cl, 1 g; NaCl, 0.8 g; $MgSO_4 \cdot 7H_2O$, 0.2 g; $CaCO_3$ (precipitated chalk), 0.2 g; KCl, 0.1 g; K_2HPO_4, 0.1 g; $CaCl_2 \cdot 2H_2O$, 0.02 g; yeast extract, 0.2 g; trace metal solution, 5 ml; vitamin solution, 10 ml; water to 1000 ml, pH, 7.0 [10, 11].

Geobacillus kaustophilus was isolated from uncooled pasteurized milk by plating on peptonized milk agar, followed by subculturing on the same medium or on nutrient agar supplemented with 1% yeast extract, 0.25% tryptophan broth, and 0.05% glucose [1]. Strains of *Geobacillus kaustophilus* are capable to grow optimally on rich media, including tryptic soy broth and Luria broth. *G. kaustophilus* A1 grown on minimal defined medium at 55°C and pH 6.5 demonstrated the ability to utilize maltose, glucose, sucrose, glycerol, fructose, galactose, citric acid, acetic acid, pyruvic acid, lactic acid, or succinic acid. Cells did not utilize dodecane, m-xylene, cellulose, oxalic acid, tartaric acid, maleic acid, propanoic acid, benzoic acid, or picolinic acid as the sole carbon sources. *G. kaustophilus* DSM7263 displayed similar characteristics, but it did not metabolize citric acid [12].

Geobacillus thermoleovorans was isolated by adding soil, mud, and water samples to L-salts basal medium supplemented with 0.1% n-heptadecane and incubated at 60°C for 1–2 weeks, followed by transfer from turbid cultures to fresh medium of the same composition. After several such transfers, pure cultures were obtained by streaking on plates with L-salts basal medium supplemented with 0.2% n-heptadecane and solidified with 2% agar [13, 14]. Strain *G. thermoleovorans* T80 displayed extremely specific glucose utilization leading to high growth rates, followed by extensive cell death and lysis with the onset of substrate exhaustion. The addition of extra carbon substrate did not halt the rapid death and lysis. Lytic phenomenon was observed for a range of different carbon substrates (glucose, pyruvate, acetate, n-hexadecane, and nutrient broth), as well as ammonium (the nitrogen source). Batch cultures grown at reduced initial substrate concentration, at lower temperatures, or at lower dilution rates than continuous-flow cultures exhibited lesser rates and degree of cell death and lysis [15]. Optical density of *G. thermoleovorans* DSM 5366 increased in casein digest medium supplemented solely with Ca^{2+} or Mg^{2+}. Na^+, and to a greater extent K^+, with concerted action of Ca^{2+} or Mg^{2+} also induced increased optical density readings of the strain [16].

Geobacillus subterraneus and *Geobacillus uzenensis* were isolated from serial dilutions of thermophilic hydrocarbon-oxidizing enrichment cultures derived from oil fields. The cultures were inoculated on agar medium supplemented with 0.1% n-hexadecane and incubated at 55–60°C [17]. *Geobacillus jurassicus* was isolated from oil field formation water by diluting enrichment cultures grown on the following medium (NH_4Cl, 1 g; KCl, 0.1 g; KH_2PO_4, 0.75 g; K_2HPO_4, 1.4 g; $MgSO_4{\cdot}7H_2O$, 0.2 g; $CaCl_2{\cdot}2H_2O$, 0.02 g; NaCl, 1.0 g; water, 1000 ml; pH 7.0) supplemented with 4% crude oil, incubated at 60°C, and plated on the same medium solidified with 2% agar [18].

Geobacillus thermocatenulatus was isolated from a slimy bloom at about 60°C on the inside surface of a pipe in a steam and gas thermal borehole using potato-peptone and meat-peptone media [19]. Some studies showed that *G. thermocatenulatus* strain does not hydrolyze starch and gelatine. It neither produces acid from xylose and lactose nor generates acetoin [20]. Trypticase soy agar can be used as the medium for cultivation of representatives of this species. Addition of 5 mg/l $MnSO_4$ encourages sporulation of the species [21].

Strains of *Geobacillus thermodenitrificans* were isolated from soil by suspending 1 g soil sample in 5 ml of sterile distilled water and heating at 90°C for 10 min, then plating 1 ml aliquot on nutrient agar and incubating at 65°C for 24 h [22]. *G. thermodenitrificans* showing L-arabinose isomerase activity was grown in media containing 1.5% pancreatic digest of casein, 0.2% yeast extract, 0.2% beef extract, 0.2% glycerol, 0.2% K_2HPO_4, 0.2% KH_2PO_4, 0.01% $MgSO_4$, 0.0004% D-biotin, and pH 6.8. Inocula were prepared by culturing the organisms for 10 h at 65°C and 200 rpm in a rotary shaking incubator using 250-ml flasks containing 50 ml of the above-described medium. These cultures were used to inoculate the fermenter where the mixture was incubated for 14 h at 65°C with agitation (400 rpm) and aeration (1.0 vvm) [23]. Some strains of *G. thermodenitrificans* were isolated and suspended in 100 ml of 0.1 mol/l potassium phosphate buffer solution (pH 7.0) with 1.0% (w/v) gelatinized potato starch in 500-ml conical flasks. The flasks were incubated at 50°C for 2 days, then 1 ml portions were transferred to 10 ml of 0.1 mol/l phosphate buffer (pH 7.0) and agitated for 6 h followed by suspension onto starch agar plates (10 g peptone, 5 g yeast extract, 10 g potato starch, and 15 g agar in 1 l of a 0.1 mol/l potassium phosphate buffer, pH 7.0) and incubation at 60°C for 24 h [24].

Geobacillus thermoglucosidasius was isolated from Japanese soil by adding 0.1 g sample to 5 ml of medium I (peptone, 5 g; meat extract, 3 g; yeast extract, 3 g; K_2HPO_4, 3 g; KH_2PO_4, 1 g; water, 1000 ml; pH 7.0) in test tubes and incubating at 65°C for 18 h, with the tubes leaning at an angle of about 10°, followed by further enrichments in tubes with the same medium and then purification on plates containing medium I solidified with 3% agar [25, 26]. Studies showed that *Geobacillus thermoglucosidasius* strains grew well and gave reproducible and comparable viable cell counts on the semi-defined agar medium (SDM) with glycerol and pyruvate as carbon sources under aerobic conditions at 70°C. SDM contains a dual carbon source (glycerol and pyruvate) and low levels of yeast extract, tryptone, and inorganic salts. The main components of the SDM are (g/l): glycerol 5.0, pyruvate 5.0, tryptone 0.2; yeast extract 0.2; citric acid 0.32; di-sodium hydrogen orthophosphate (anhydrous) 2.0; magnesium sulfate (heptahydrate) 0.4; potassium sulfate 0.3; ammonium chloride 2.0; manganese chloride (tetrahydrate) 0.003; ferric chloride 0.007; agar 15, and 1 ml of trace elements solution. SDM formulation suggests joint preparation of glycerol and pyruvate separated from the other constituents and mixing after sterilization at 121°C [27].

Geobacillus lituanicus was isolated using 10-fold serial dilutions of crude oil. The dilutions were inoculated onto Czapek agar and plates were incubated aerobically at 60°C for 48 h [28]. *Geobacillus toebii* was isolated from a suspension of hay compost plated onto solid modified basal medium (polypeptone, 5 g; K_2HPO_4, 6 g; KH_2PO_4, 2 g; yeast extract, 1 g; $MgSO_4{\cdot}7H_2O$, 0.5 g; L-tyrosine, 0.5 g; agar to solidify; and deionized water, 1000 ml) and incubated at 60°C for 3 days [29]. *G. toebii* subsp. *decanicus* was cultivated in medium containing 8 g/l peptone, 4 g/l yeast extract, 2 g/l NaCl at pH 7.0 [30]. *Geobacillus vulcani* was isolated from a marine sediment sample by inoculating into Bacto marine broth and medium D and incubating aerobically for 3 days at 65°C, followed by plating positive cultures again onto Bacto marine agar [31]. *Geobacillus galactosidasius* was isolated from a compost sample and inoculated in 10 ml of enrichment media cultured within pH range 5.5–7.2 and the temperature range 60–80°C for 3 days. The enrichment growth media used were medium A containing 8 g/l peptone, 4 g/l yeast extract, and 2 g/l NaCl at pH 7.2, and medium B containing 6 g/l yeast extract and 6 g/l NaCl at pH 5.5 [32]. *Geobacillus icigianus* was isolated from sludge samples of an explosive hydrothermal spring located near the Troinoy geyser and purified on LB agar medium at 55–65°C [33].

3. *Geobacillus* bacteria: taxonomy and identification

Thermophilic bacterium known as *Bacillus stearothermophilus* (now typical species *Geobacillus stearothermophilus*) was discovered in 1920 [34]. For many years, geobacilli have been referred to *Bacillus* species. The development of molecular genetic methods resulted in the division of bacilli into several phylogenetically distinct genera. Group 5 including *B. stearothermophilus*, *Bacillus kaustophilus*, and *Bacillus thermoglucosidasius* formed a generic lineage distinct from *Bacillus* species [35]. In 2001, Nazina et al. proposed that the six species of bacilli, namely *B. stearothermophilus*, *B. kaustophilus*, *B. thermoglucosidasius*, *B. thermocatenulatus*, *B. thermoleovorans*, and *B. thermodenitrificans*, should be incorporated into a new genus, *Geobacillus*, along with two novel species, *Geobacillus subterraneus* and *Geobacillus uzenensis* based on 16S rRNA gene sequence analysis and a variety of physical and biochemical characteristics [17].

At present, geobacilli include about 20 species [36]. However, the lowest level of 16S rRNA gene sequence similarity between all *Geobacillus* species is around 93%, which indicates that at least some species need to be reclassified at the genus level [1]. Full-length *recN* and 16S rRNA gene sequence analysis clustered 68 isolates from the genus *Geobacillus* into 9 similarity groups. Some of these groups corresponded unambiguously to the known species, whereas the others contained two or more type strains from species with validly published names. *recN* was clearly superior to the 16S rRNA gene, with nearly an order of magnitude greater resolving power at the species-subspecies level, but the analysis was much less effective for higher taxa [37]. The availability of complete genome sequence data offers higher resolution of the phylogenetic relationships of *Geobacillus* species than the use of a single housekeeping gene. For example, the two strains of *G. kaustophilus* do not form a phylogenetically coherent monophyletic clade, while the two strains of *G. thermoleovorans* are closely related and share 99.4% nucleotide sequence identity [38]. The use of the phylogenomic metrics such as average amino acid identity (AAI), average nucleotide identity (ANI), and digital DNA-DNA hybridization (dDDH) indicated that the current genus *Geobacillus* is composed of 16 distinct genomospecies. Phylogeny constructed on the basis of the core genes demonstrated that the genus clustered into two monophyletic clades differing in terms of nucleotide base composition. The G + C content ranges for clade I and II were 48.8–53.1 and 42.1–44.4%, respectively. The *Geobacillus* species within clade II can be considered as a new genus [39].

It was shown that evolutionary conservatism of 16S rRNA leads to the case, when different bacterial strains belonging to the same species or closely related species may have identical sequences. It decreases efficiency of 16S rRNA analysis and provokes the search for alternative approaches, like comparative analysis of the nucleotide sequences of "housekeeping genes" determining the main metabolic processes. These genes have some advantages similar to those of 16S rRNA genes: universal distribution, evolutionary conservatism, and vertical inheritance. At the same time, comparative phylogenetic analysis of "housekeeping genes" may be more efficient at the lowest taxonomic levels for the following reasons: (1) the level of conservatism varies for different genes and distinctions between the nucleotide sequences may be more pronounced as compared to sequences of 16S rRNA; (2) "housekeeping genes" code for proteins allowing to better determine the frequency of synonymous substitutions and locate taxonomic position; (3) the majority of these genes are represented by a single copy. Besides the abovementioned *recN* gene, application prospects have been demonstrated for other genes. Genes encoding the β-subunits of a type II topoisomerase (gyrase, *gyrB*) and a type IV topoisomerase (*parE*) provided for a more precise determination of the phylogenetic position of bacteria at the species level as compared to 16S rRNA analysis of *Geobacillus* [40]. Additionally, genes *recA* and *rpoB* can be used in identification and taxonomic affiliation of *Geobacillus* species [41]. Gene *spo0A* codes for protein serving as the master regulator of the endospore formation process in the endosporulating bacteria and can be engaged in taxonomic positioning of these bacterial groups. The gene cannot be used as the phylogenetic marker within the genus *Geobacillus*, although it was shown to be helpful for the identification of *G. thermodenitrificans*, *G. stearothermophilus*, *G. jurassicus*, and cluster *G. subterraneus*-*G. uzenensis* [42].

16S-23S rRNA gene internal transcribed spacer (ITS) separates 16S and 23S rRNA genes and may contain tRNA genes. ITS exhibits larger variations in both the length and the sequence of this region and can be more useful in identification than 16S rRNA. *Geobacillus* genus-specific primers GEOBAC allowed to amplify the DNA from 13 species. Different species could be grouped according to the number and size of the PCR products and identified on the basis of the restriction analysis of these products [43].

Repetitive DNA is also a promising tool for identification of bacteria. Usually, repetitive DNA is applied for genotyping of medically important microorganisms, but it may be employed in analysis of biotechnologically important *Geobacillus* species. In total, 33 *Geobacillus* genus-specific motifs with length over 20 nucleotides were determined: 3 were genus-specific, 15—species-specific, and 15—species cluster-specific. Motifs have been used for the construction of the genus- and species-specific (*G. thermodenitrificans* and *G. toebii*) primer pairs [44]. Rep-PCR molecular method based on the usage of outwardly facing oligonucleotide PCR primers complementary to interspersed repetitive sequences is widely applied in the characterization of different groups of bacteria. BOX, ERIC, REP, and $(GTG)_5$ are examples of evolutionarily conserved repetitive sequences. Rep-PCR fingerprinting technique using $(GTG)_5$- and BOX-PCR has been shown as a rapid, easy-to-conduct, and reproducible tool for differentiation of thermophilic bacteria at the species, subspecies, and potentially up to the strain level, with a single-performance protocol [45]. Multilocus variable-number tandem-repeat (VNTR) analysis (MLVA) has been used to discriminate between different genotypes within species by analyzing length polymorphism of several VNTR loci, while high-resolution melt analysis (HRMA) has been shown as post-PCR method for analysis of genetic variations in PCR products using DNA-binding fluorescent dyes and a PCR machine with a highly precise temperature control. HRMA in conjunction with MLVA (MLV-HRMA) displayed a stronger discriminatory power and better reproducibility than RAPD-PCR and hence can be used for genotyping *Geobacillus* species [46].

Amplified ribosomal DNA restriction analysis (ARDRA) using *AluI* was shown to be a valuable, easy, and accurate technique for the identification of *G. stearothermophilus*. The presence of a fragment 162 bp in size and the absence of 76 and 86 bp fragments were recognized to be characteristic traits of this species. However, the potential of the method for the identification of other species of geobacilli is limited [47].

4. Applications of *Geobacillus* bacteria

Geobacillus species are represented by thermophilic chemoorganotrophs. They are able to utilize a wide range of substrates, including hydrocarbons, and produce various metabolites of commercial use, like enzymes (proteases, lipases, and amylases), ethanol, bacteriocins, etc. Such metabolic diversity coupled to high temperature resistance makes the bacteria attractive for various applications. On the other hand, *Geobacillus* species are known to be a major cause of spoilage in canned food. The use of thermostable enzymes and biofuel production are considered in the next chapters.

Petroleum is a complex compound consisting of hydrocarbons, a small ratio of nonhydrocarbon components, and trace metals. In order to increase oil production, traditional water flooding method is used. Heavy oil, the largest potentially recoverable petroleum energy resource, is very viscous and has a high freezing point. It is problematic to extract the residual crude oil by the conventional method, so that alternative techniques, like use of microorganisms, for enhanced oil recovery have been proposed. Microorganisms and their metabolites help to retrieve residual oil by promoting its emulsification and reducing viscosity. Since high temperature reservoirs are more difficult to exploit by this method, thermophilic bacteria are advantageous in this case. *G. stearothermophilus* A-2 shows strong surface hydrophobicity and produces a bioemulsifier. The fermentation broth of strain A-2 induced crude oil dispersion and decreased oil viscosity. Moreover, strain A-2 preferentially degraded heavy oil components and polycyclic aromatic hydrocarbons. These features make the strain an excellent candidate for enhanced microbial oil recovery from high-temperature deposits [48]. *Geobacillus* sp. ZY-10 could utilize tridecane, hexadecane, octacosane, and hexatridecane as the sole carbon sources, and the digestion rate of long-chain alkanes was lower than that of short-chain alkanes. Addition of inorganic salts and trace yeast extract led to the significantly increased concentration of short-chain alkanes and the decreased content of long-chain alkanes, suggesting that the larger hydrocarbon components in crude oil were converted into shorter-chain alkanes. Thus, strain ZY-10 proved effective for improving the mobility and upgrading quality of heavy crude oil [49]. *G. pallidus* H9 was able to grow in temperature range 45–80°C at salinity 0–15% and synthesize biosurfactant using crude oil as the sole carbon source under aerobic or anaerobic conditions. After incubation in LB medium, 20 ml bacterial suspension was transferred to 500 ml triangular flask with 10 g of sterile crude oil-containing mineral medium, and incubated at 65°C and 180 rpm for 100 days under aerobic or anaerobic conditions. The medium was boiled for 30 min in order to expel all dissolved oxygen prior to incubation under anaerobic conditions; L-cysteine and resazurin as oxygen indicators were added to the medium to final concentrations of 0.05% and 0.01 (g/l), respectively. The yields of biosurfactant were ≈9.8 and ≈2.8 g/l under aerobic and anaerobic conditions, respectively. The fractionated components and composition of the purified biosurfactant differed between aerobic (glycosides≈50.3%, lipids≈34.5%, and peptides≈15.2%, w/w) and anaerobic (glycosides≈53.8%, lipids≈31.2%, and peptides≈26.0%, w/w) cultures. The maximum production of biosurfactant under aerobic conditions is determined by the different electron acceptors, resulting in the different pathways of biodegradation with different reaction activation energy and generation of different catabolic enzymes. The strain H9 showed preference for utilization of medium- and long-chain alkanes (C23-C43) under aerobic conditions, and degradation of long alkanes (C33-C43) under anaerobic conditions. The evident difference in the metabolic pathways between aerobic and anaerobic degradation was possibly due to a change in redox potential during the biochemical reaction. The strain H9 and its biosurfactant are potentially promising agents intensifying microbial oil recovery, especially in high temperature and salinity oil reservoirs [50]. Besides, the ability to utilize hydrocarbons makes geobacilli indispensable in bioremediation of environment, like removal of oil spills [51].

Bacteriocins are ribosomally synthesized antimicrobial peptides or proteins, usually possessing the narrow antagonistic activity spectrum against bacterial strains closely related to the strain-producer. The activity of this group of molecules against foodborne and pathogenic bacteria opens wide opportunities for their application in medicine and food industry. Little is known about bacteriocins of thermophilic bacteria; however, *Geobacillus* species demonstrated antibacterial activity. *Geobacillus* strains isolated from the surface soil above oil deposits were active against at least 1 of 19 tested pathogenic bacteria. The derived antibacterial compounds were stable in broad temperature and pH ranges, sensitive to proteolytic enzymes, proving their proteinaceous nature. They were active against closely related thermophilic bacteria, which suggests that these substances are most likely bacteriocins [52]. *Geobacillus* species have been shown to produce antimicrobials such as antibiotics. *G. thermodenitrificans* NG80-2 synthesizes two antibiotics: geobacillin I and II. The former antibiotic is nisin analog showing antimicrobial spectrum similar to nisin A, with increased activity against *Streptococcus dysgalactiae*, one of the causative agents of bovine mastitis. Geobacillin I demonstrated increased stability compared to nisin A. Geobacillin II displayed antimicrobial activity only against *Bacillus* strains [53].

Nanomaterials are defined as engineered materials with a least one dimension in the range of 1–100 nm. They often exhibit unique and considerably modified physical, chemical, and biological properties. Nanomaterials are used in cosmetics, chemical, and food industries, construction, medicine, agriculture, production of electronic equipment and sensors, etc. [54, 55]. The available physical methods for the metal nanoparticle synthesis such as gas condensation and irradiation with ultraviolet or gamma rays usually resulted in low production rate and high expenditure. The large scale synthesis of metal nanomaterials suffers from certain drawbacks such as polydispersity and lack of stability, especially if the reduction is carried out in aqueous media. Biological synthesis of nanoparticles can be an alternative choice. The exposure of *G. stearothermophilus* cell-free extract to the metal salts leads to the formation of stable silver and gold nanoparticles in the solution. The stability of nanoparticle solution could be due to the secretion of certain reducing enzymes and capping proteins by the bacterium. Preliminary gel electrophoresis indicates that the bacterium secretes not less than seven different proteins of molecular mass ranging between 12 and 98 kDa. One or more of these proteins could be a reductase enzyme that reduces metal ions [56]. Silver nanoparticles can be produced from silver nitrate using spore extract of *G. stearothermophilus*. Cytotoxicity of nanoparticles derived from this extract toward microbial pathogens such as *Candida albicans*, *Candida glabrata*, *Streptococcus mutans*, and *Streptococcus sobrinus* was generally higher than cytotoxic effect of silver nanoparticles stemming from spore extract of *Bacillus subtilis*. Moreover, antibacterial effects significantly surpassed antifungal effects [57].

Various human activities generate wastewater containing nitrogenous compounds. In the natural environment, nitrogen removal often occurs at 20–30°C, temperatures suitable for growth of nitrifying and denitrifying bacteria. As a result, many wastewater treatment plants use mesophilic bacteria for bioremediation. Thermophilic bacteria are expected to have increased enzyme activity and stability in comparison with mesophilic ones in such applications. The thermophilic denitrifying bacterium *Geobacillus* sp. strain TDN01 showed 12 times higher specific nitrate removal rate on media with ammonia than without ammonia. The consumption rates of nitrate and succinate were proportional. The growth rates with 120 and 150 mM nitrate were only slightly lower than those with 60 mM, not leading to notable growth inhibition. The maximum denitrification rate was six times higher than that of mesophilic bacteria [58]. Organophosphonates are characterized by the presence of a stable, covalent C—P bond. One of important applications for synthetic organophosphonates is manufacturing of herbicides, such as glyphosate. Glyphosate is the most widely used nonselective herbicide worldwide. It can cause a wide range of clinical manifestations in human beings, like skin and throat irritation to hypotension, oliguria, and death [59]. Strain *G. caldoxylosilyticus* T20 could utilize a number of organophosphonates as the sole phosphorus source for growth at 60°C. During growth on glyphosate, aminomethylphosphonate release to the medium was observed [60]. Azo dyes are characterized by the presence of one or more azo bonds (—N=N—) in association with one or more aromatic systems, which may carry sulfonic acid groups. The pigments are extensively used in the dyeing and textile industries and can provoke grave problems when discharged in the environment. *G. stearothermophilus* UCP 986 was able to degrade 96–98% of the azo dye after 24 h of incubation on LB medium under aeration. The brine shrimp *Artemia salina* showed the absence of toxic metabolites during the decolorization process. However, increased concentration of the dye and vigorous agitation led to high mortality rate of the shrimp [61]. Synthetic polymers generally display strong resistance to biodegradation, causing serious pollution problems as wastes persisting in the environment for a long time. Aliphatic polyamides (nylons), like nylon 6, nylon 66, and nylon 12, are produced in large amounts and are regarded as recalcitrant to biodegradation. At 60°C, *G. thermocatenulatus* grew on 5 g/l nylon 12, decreasing its molecular weight from 41,000 to 11,000 over 20 days. The strain also degraded nylon 66 with a decline in its molecular weight from 43,000 to 17,000 in 20 days at 60°C. However, nylon 6 was not utilized [62].

The heavy metals are well-known toxicants and their determination is vital for ecological control of soils, food, and water. Cells of *G. thermoleovorans* subsp. *stromboliensis*, immobilized on Amberlite XAD-4, showed the sorption capacity of 0.0373 and 0.0557 mmol/g for Cd(II) and Ni(II), respectively. The detection limits were 0.24 μg/l for cadmium and 0.3 μg/l for nickel. The system sustained 20 operation cycles without any loss in its sorption potential and can be used for metal determination in water and food samples [63]. Bacteria may be also applied in biosorption and removal of toxic metals from aqueous solutions. Dead biomass of *G. thermodenitrificans* MTCC 8341 in the synthetic metal solutions reduced the concentration of Fe^{+3} (91.31%), Cr^{+3} (80.80%), Co^{+2} (79.71%), Cu^{+2} (57.14%), Zn^{+2} (55.14%), Cd^{+2} (49.02%), Ag^{+} (43.25%), and Pb^{+2} (36.86%) at different optimum pH values within 720 min. The strain in the industrial wastewater reduced concentrations to 43.94% for Fe^{+3}, 39.2% for Cr^{+3}, 35.88% for Cd^{+2}, 18.22% for Pb^{+2}, 13.03% for Cu^{+2}, 11.43% for Co^{+2}, 9.02% for Zn^{+2}, and 7.65% for Ag^{+} within 120 min. [64]. In turn, Cd^{2+}, Cu^{2+}, Co^{2+}, and Mn^{2+} removal at 50 mg/l concentration in 60 min by 50 mg of dry *G. thermantarcticus* cells at optimum growth temperatures was 85.4, 46.3, 43.6, and 65.1%, respectively [65]. Thus, *Geobacillus* strains may be used for disposal of heavy metals.

Thermophilic bacteria can be the sources of valuable biomolecules, like complex polysaccharides for medical application. Extracellular polysaccharide produced by strain of *G. thermodenitrificans* hinders HSV-2 replication in human peripheral blood mononuclear cells (PBMC). High levels of IFN-α, IL-12, IFN-γ, TNF-α, and IL-18 were detected in exopolysaccharide supernatants following PBMC treatment. This effect was dose-dependent. Therefore, the immunological disorders determined by HSV-2 could be partially resolved by treatment with polysaccharide [66].

The biosurfactants, amphiphilic molecules consisting of hydrophobic and hydrophilic groups, are used in cosmetic formulas, pharmaceutical, chemical, food industries, agriculture, production of cleansers, enhanced oil recovery, and in bioremediation of oil-contaminated sites. Biosurfactant of *G. stearothermophilus* UCP 0986 grown on corn steep liquor and palm oil medium is capable of reducing the surface and interfacial water tensions to significantly lower values, and possesses excellent emulsifying and dispersion properties. The biosurfactant was isolated by precipitation of metabolic cell-free liquid with acetone 1:1 (v/v). The precipitate was allowed to stand for 24 h at 4°C, and then, it was centrifuged at 4000 rpm for 15 min at 5°C. The supernatant was discarded and the isolated biosurfactant was subjected to dialysis against deionized water for 72 h at 5°C. Properties of the compound arouse interest in terms of bioremediation of hydrophobic molecules [67].

Lactic acid is used in several industrial sectors producing food, pharmaceuticals, chemicals, and cosmetics. Around 90% of its entire production is derived from fermentation by lactic acid bacteria. However, application of thermophilic bacteria may reduce the risk of contamination and sterilization cost. *G. stearothermophilus* DSM494 generated lactic acid as the major product of anaerobic metabolism. The strain produced 37 g/l optically pure (98%) L-lactic acid in 20 h from 50 g/l raw potato starch. Smaller amounts (<7%w/v) of acetate, formate, and ethanol were also formed. Yields of lactic acid increased from 66 to 81% when potato residues from food processing were used as a starchy substrate in place of raw potato starch [68].

Non-standard application for *Geobacillus* has been demonstrated in the enhancement of cement-sand mortar properties. The *G. stearothermophilus* was chosen for the ability to resist extreme environmental factors. Bacteria at concentration 1×10^9 CFU/ml improved the performance of cement-sand mortar in terms of compressive strength and water absorption as compared to the control mortar. The incorporation of *G. stearothermophilus* has catalyzed the occurrence of wollastonite ($CaSiO_3$) capable to upgrade concrete properties by modifying its pore structure [69].

Geobacillus species can be used for cultivation and study of other bacteria. Many microorganisms in nature cannot be successfully cultured under artificial conditions, even in the presence of appropriate nutrients. *Symbiobacterium toebii* requires some growth-supporting factors from its partner bacterium *G. toebii* and does not show sustainable growth in artificial culture in the absence of the partner or its supernatant/cell-free extract [70]. *Geobacillus* sp. W2-10 enhances the cellulose-degrading activity of cellulolytic bacteria *Clostridium thermocellum* CTL-6 in peptone-cellulose solution medium under aerobic conditions. Cellulose degradation efficiency of filter paper and alkaline-treated wheat straw significantly increased up to 72.45 and 37.79%, respectively. The carboxymethyl cellulase activity and biomass productively of CTL-6 also rose from 0.23 U/ml and 45.1 μg/ml up to 0.47 U/ml and 112.2 μg/ml, respectively. In addition, coculture resulted in the accumulation of acetate and propionate up to 4.26 and 2.76 mg/ml [71].

Geobacillus species are able to reduce effects of osmotic stress in plants. *G. caldoxylosilyticus* IRD, halophilic facultative aerobic bacterium, inoculated into 5-day-old maize cultivars prior to treatment with 350 mM NaCl for 10 days improved maize growth and dry weight. The number of vascular bundles decreased in roots and increased in leaves upon inoculation with bacteria. In addition, the accumulation of toxic Na^+ and Cl^- was much lower in treated seedlings. Proline level, stress indicator, became two to four times higher in seedlings exposed to salt without *Geobacillus* [72].

Crystalline bacterial cell surface layers (S-layers) are composed of identical proteins or glycoprotein subunits which can self-assemble into two-dimensional crystalline arrays. It has been shown that S-layers of *Geobacillus* could be used as a biological template for immobilization of molecular array and provide new approaches for nanoelectronic biosensor design [73].

Probiotics are health-promoting microbial agents. Mechanisms of probiotic action are diverse. Positive effect can be expressed as direct action on the organism, like enhancement of barrier function, metabolism, immunomodulatory effects, or action on pathogenic microorganisms via secretion of bioactive compounds and/or competition with pathogens (**Figure 2**). The global probiotics market reached $31.8 and $34.0 billion in 2014 and 2015, respectively. The market capacity is likely to expand to $50.0 billion by 2020, growing at compound annual growth rate of 8.0% from 2015 to 2020 (**Figure 3**) [74]. Lactic acid bacteria are main sources of probiotics. However, it was reported that a number of bacteria such us *Bacillus* spp., *Aeromonas hydrophila*, and yeasts such us *Saccharomyces boulardii*, *Candida famata*, and *Candida parapsilosis* possess probiotic potential. *Geobacillus* species are not usually regarded as probiotic candidates, but these bacteria are able to produce a number of beneficial compounds that can favorably influence the organism or demonstrate antimicrobial features. Some bacteria show the adherence ability. Thus, *Geobacillus* species can be considered as potential sources of probiotics [75].

Due to their resistance and spore formation, geobacilli may be used as biological indicators of sterility. They are usually applied in the studies on efficiency of sterilization processes [76–78]. Finally, *Geobacillus* species can find use in engineering applications, such as directed evolution of robust variants of mesophilic proteins [79].

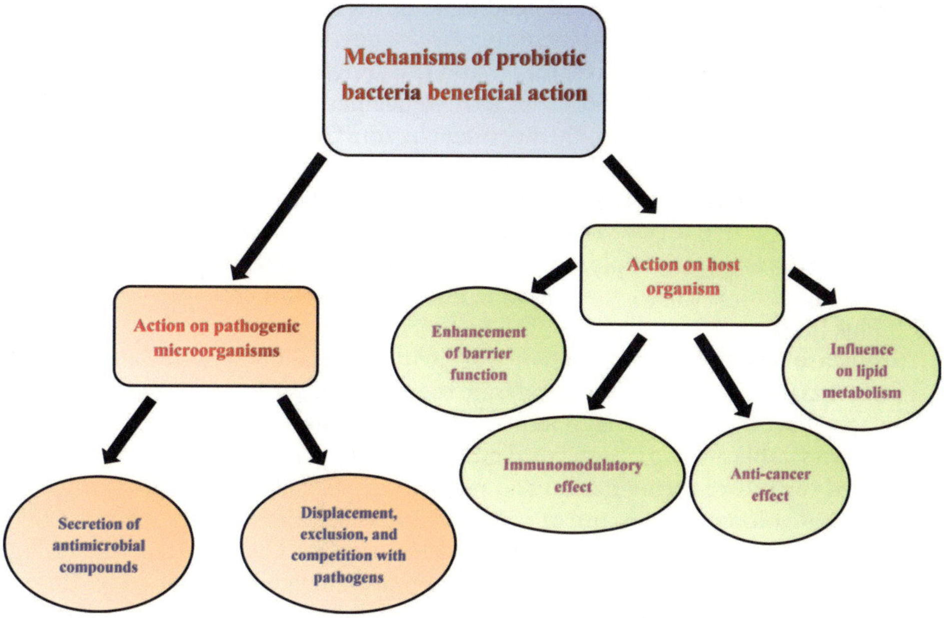

Figure 2.
Mechanisms of beneficial action of probiotic bacteria.

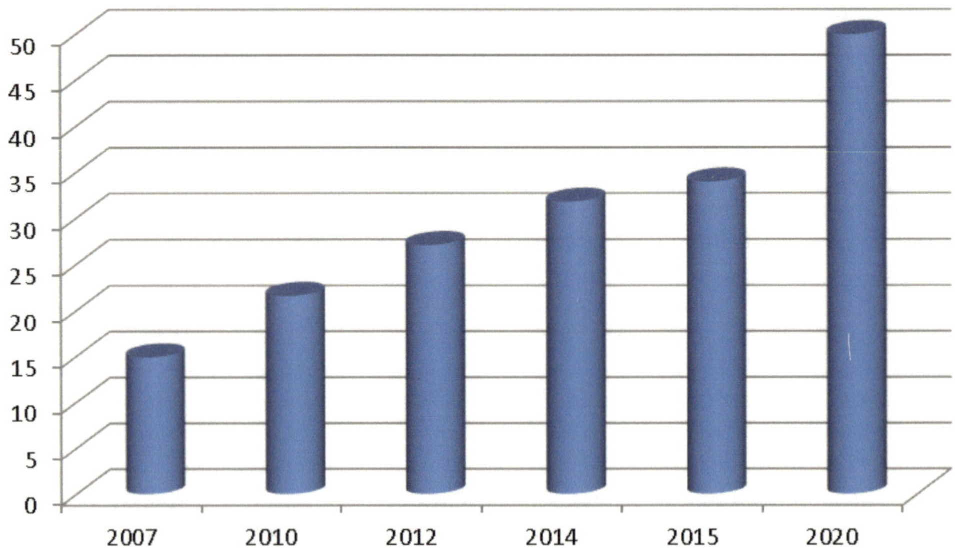

Figure 3.
Probiotics market (USD billion dollars) growth (years 2007–2020).

5. Production and applications of thermostable enzymes of *Geobacillus* bacteria

Pectinases or pectinolytic enzymes hydrolyze pectic substances. Pectinolytic enzymes are involved in fruit juice extraction and clarification, scouring of cotton, degumming of plant fibers, wastewater treatment, vegetable oil extraction, tea and coffee fermentations, bleaching of paper, in fabrication of poultry feed additives and alcoholic beverages, and in food industries [80]. The raw pressed juice is rich in insoluble particles mainly made up by pectic substances. It is difficult to extract juice by pressing or using other mechanical methods. The addition of pectinases decreases viscosity of the fruit juice, leading to its higher yields. These enzymes are produced by numerous bacteria, including *Geobacillus* species. The purified pectin lyase of *G. pallidus* P26 retained stability and full activity after 24 h incubation in temperature range from 40 to 50°C. Its activity decreased when the temperature increased above 70°C. However, the purified enzyme from thermophilic bacteria can keep its activity for 2 and 5 h at 80 and 70°C, respectively. In addition, the activity fell only by 50% after 24 h at 60°C. The optimal pH value for pectin lyase was 9, and it remained active between pH 5 and 11. Compared to the control, the enzyme increased juice yield from apple, banana, carrot, and peach pulps. Nevertheless, the results obtained for purified enzyme were less than those for commercial Pectinex 100 L Plus [81].

α-Amylase is one of the enzymes of worldwide interest in food, pharmaceutical, and fermentation industries. The moderate thermostability and Ca^{2+} requirement of α-amylases limit their industrial potential. The supplementation of wheat flour with hyperthermostable and Ca^{2+}-independent α-amylase of *G. thermoleovorans* NP54 accelerated the rate of fermentation and reduced the viscosity of dough, resulting in the improved volume and texture of bread, its increased shelf life and softness [82]. The same amylase used in starch saccharification produced hydrolysate containing a high proportion of maltose. As a consequence, this enzyme can find use in the manufacture of high maltose syrups consumed by the food, chemical, and pharmaceutical industries [83]. The amylase from *Geobacillus* sp. IIPTN was stable over a broad range of temperatures from 40 to 120°C and pH from 5 to 10 and showed resistance to protease. These characteristics emphasize the enzyme potential in industrial applications [84]. Engineered α-amylase of *G. stearothermophilus* US100 with its significantly lower requirement for calcium ions, high resistance to thermal inactivation, to chelators, to protease, and to oxidative additives, in conjunction with well-preserved activity after storage, may be an excellent candidate for manufacturing enzyme detergent [85].

α-Glucosidases are hydrolases releasing α-glucose from the ends of the substrates such as oligo- and polysaccharides. They are usually found in association with other amylolytic enzymes which accomplish complete degradation of starch. α-Glucosidases show diversity in substrate specificity and transglycosylation activities, and such specificity differs considerably with the enzyme source. They have a number of potential applications in fundamental research, industrial starch processes, α-amylase assay kits for clinical laboratories, and synthesis of oligo-, di-, and trisaccharides. The thermostable exo-α-1,4-glucosidase of strain G. *thermodenitrificans* F84a remained active over temperature range 35–70°C and pH range 4.5–11.0, with optimum activity at 60°C and pH 7.0. The α-glucosidase hydrolyzed α-1,6, α-1,3, and α-1,4 bonds of substrate molecules in addition to a high transglycosylation activity. The enzyme was also found to be resistant to most of the denaturing agents and inhibitors. The characteristics of α-glucosidase make it a promising agent for biotechnological processes. It can be used for enzymatic synthesis of novel tri- and oligosaccharides due to high conformational stability and transglycosylation activity [86]. α-Glucosidase from *Geobacillus* sp. strain HTA-462 isolated from sediment of the Mariana Trench exclusively hydrolyzed α-1,4-glycosidic linkages of oligosaccharides in an exo-type manner. The enzyme showed an overwhelming transglycosylation activity and glycosylated various nonsugar molecules when maltose was used as a sugar donor. It converted maltose to isomaltose. α-Glucosidase is also a potential catalyst in the biosynthesis of complex carbohydrates [87].

Cellulose is the major component of plant biomass. The enzymes degrading it are applied in the textile industry for cotton softening, in the production of detergents for color care and cleaning, in the food industry for mashing, and in the pulp and paper industries for deinking, drainage promotion, and fiber modification. Thermophilic cellulases show advantages in many industrial applications because elevated processing temperatures provide for accelerated reaction rates, increased solubility of reagents and reduced contamination. Endoglucanase of *Geobacillus* sp. 70PC53 expressed an optimal activity at 65°C and pH 5.0, and it exhibited 10-fold higher specific activity than the commercially available *Trichoderma reesei* endoglucanase. The enzyme displayed activity over a broad temperature range from 45 to 75°C and good prospects in biomass conversion, detergent upgrading, paper pulping, textile manufacturing, and juice clarification [88]. The thermophilic *Geobacillus* sp. T1 is able to grow and produce cellulase efficiently on untreated wheat and barley straw as the sole carbon sources. It harbors the potential for conversion of agricultural biomass to fuels [89]. Endocellulase of *G. thermoleovorans* T4 can hydrolyze carboxymethylcellulose, phosphoric acid-swollen cellulose, Avicel, filter paper, and salicin. When the strain was grown in medium with carboxymethylcellulose, the cellulolytic enzyme activity in culture supernatants was stable up to 70°C. More than 10% of the original activity was still detectable after heating to 100°C at pH 7.0 for 1 h [90].

Lipase catalyzes hydrolysis of triglycerides and produces esters by esterification reaction. They are used in many sectors such as food, pharmaceutical, chemical, petrochemical, biodiesel, and in detergent industries. Their main application is the enantioselective synthesis of precursors of pharmaceutically active compounds and the conversion of natural fats and oils into high-value products such as cocoa, butter, and oil enriched with omega-3 fatty acids. Thermostable lipases are characterized by inherent stability, so that they can find use in various industries and biotechnological sectors as additives to detergents, food processing aids, environmental bioremediation agents, and molecular biology tools. Lipase from *Geobacillus* sp. immobilized by surface adsorption onto silica showed maximum activity at temperature 55°C and 82.94% yield of methyl salicylate. The latter has various medicinal applications, like control of muscular pain [91]. In leather industry, the addition of hydrolytic enzymes such as lipases and proteases in the soaking step facilitates fat degradation and raises leather quality. Thermoalkaliphilic lipase from *G. thermoleovorans* DA2 is produced in high amounts using cheap substrate, such as fatty restaurant wastes, thus making the production process cost-effective. The lipase with Triton X-100 proved the best degreasing agent by lowering the total lipid content to 2.6% as compared to kerosene, commonly utilized organic solvent, (7.5%) or the sole crude enzyme (8.9%). As a result, the chemical leather process can be substituted with thermoalkaliphilic lipase treatment to upgrade the quality of leather and reduce the environmental hazards [92]. Thermostable lipase from *G. thermodenitrificans* IBRL-nra exhibited the highest stability in the presence of acetone, ethanol, and acetonitrile and showed elevated activity (220%) when pretreated with Triton X-100. It could preserve 100% of its activity in the presence of protease up to 4 h and could retain 70% of its initial activity after 24 h of incubation; hence, the enzyme can be applied in biotechnological processes and industries [93]. Lipase from *G. thermodenitrificans* nr68 has expressed great enzymatic biodeinking activity toward a laser jet printed paper, with deinkability brightness test of 55%, a value that was slightly lower than that shown by a commercial lipase (63%). Such lipases can be used in the recycling of waste paper [94]. *Geobacillus* sp. T1 lipase fused with a cellulose binding domain has a strong medium chain (C8:0 and C10:0) preference and acts weakly on C18:1 n-9 in acidolysis reactions. This enzyme could be used as a potential biocatalyst in the synthesis of structured lipids. The latter are triacylglycerols carrying particular fatty acids in certain positions of the glycerol backbone, which provide for nutritive or therapeutic purposes [95].

Pullulanases are debranching enzymes able to hydrolyze the alpha-1,6 glycosidic linkage in pullulan, starch, amylopectin, and related oligosaccharides. Pullulanase is used for the production of glucose or maltose syrups in a two-stage liquefaction and saccharification process. Temperature variation in the saccharification process causes deactivation of enzymes and therefore the increase of enzyme costs, so that application of thermostable pullulanase could decrease the industrial costs and increase the process efficiency by merging the two-step liquefaction and saccharification scheme into one stage. The enzyme from strain *Geobacillus* sp. LM14-3 showed maximum activity at 60°C and pH 6.5 with a half-life time about 56 h, which favors the potential industrial application [96].

Xylan is a major component of hemicellulose. Its complete hydrolysis requires the combined action of various enzymes, like endo-1,4-β-xylanase, β-xylosidase, α-arabinofuranosidase, acetylxylan esterase, α-glucuronidase, and feruloyl esterase. Applications of xylanases include food, pulp and paper industry, and agriculture. The great majority of these enzymes are optimally active in the acidic or neutral pH range at temperatures up to 60°C. Since the incoming pulp for enzymatic bleaching is hot and alkaline, the use of thermostable alkaline xylanases is very attractive from economical and technical point of view. *Geobacillus* sp. 71 produces xylanase at optimum temperature 75°C and pH 8.0, but it is active over a broad pH range. The action of the enzyme on oat spelt xylan produced xylobiose and xylotetrose. Due to its characteristics, the enzyme is a promising candidate for the production of xylosaccharides in the pulp and paper and food industries [97]. Xylanase of *G. thermoleovorans* was optimally active at pH 8.5 and 80°C, and it was found to be useful in the pulp prebleaching process, since cellulases may adversely affect the quality of the paper pulp by destroying the structure of cellulose [98]. The immobilized xylanase from thermophilic *Geobacillus* sp. TF16 displayed an increase in optimum temperature from 55 to 65°C and shift in the pH value from 6.0 to 8.5 as compared to the free enzyme. The enzyme was found to be effective in release of the reducing sugars from juice and poultry feed and oven spring in bakery [99]. The β-xylosidase of the extremely thermophilic *G. thermodenitrificans* is optimally active at 60°C and pH 7.0 and it catalyzes transxylosylation reactions in the presence of alcohols as acceptors. The pharmaceutically important β-methyl-D-xylosides could be produced using β-xyloside as the donor and methanol as acceptor [100]. L-Arabinosyl residues widely distributed in hemicelluloses constitute monomeric and/or oligomeric side chains on the β-(1 → 4)-linked xylose or galactose backbones in xylans, arabinoxylans, and arabinogalactans and make up the core in arabinans forming α-(1 → 5)-linkages. α-L-Arabinofuranosidases are enzymes hydrolyzing terminal nonreducing α-L-arabinofuranosyl groups in L-arabinose-containing polysaccharides. α-L-Arabinofuranosidase from *G. caldoxylolyticus* TK4 released L-arabinose from arabinan and arabinooligosaccharides. No endoarabinanase activity was detected. L-Arabinose has a sweet taste, is not readily assimilated by the body, hence it can be used as a food additive [101].

Gellan synthesized by *Sphingomonas paucimobilis* is the microbial exopolysaccharide finding recently extensive use in food, pharmaceutical industries and in microbial cultivation media. Gellan lyase lowers the gellan viscosity in solutions and might broaden its current spectrum of application. However, gellan is soluble at temperatures higher than 50°C, so that industries need a thermostable gellan lyase. The enzyme from *G. stearothermophilus* 98 demonstrated maximum activity at 70°C. The thermal denaturation curve of the enzyme at 214 nm showed a highly cooperative transition with a midpoint at about 75°C. Kinetic studies indicated high affinity of the enzyme to gellan as a substrate. A single product was identified after enzyme action on gellan [102].

β-Galactosidases catalyze hydrolysis of lactose into glucose and galactose and take part in transgalactosylation reaction that produces galactooligosaccharide. β-Galactosidase has been used in biopharmaceutical, food, and dairy industries to prevent crystallization of lactose, to improve sweetness, to increase the solubility of milk products, to prepare low lactose-containing food products for relatively lactose-intolerant people, and to utilize cheese whey, which would otherwise become an environmental pollutant [103]. Some *Geobacillus* species possess these enzymes. Evolved β-galactosidases BgaB from *G. stearothermophilus* KVE39 provide for transglycosylation of lactose into oligosaccharide characterized as 3′-galactosyl-lactose. Galactooligosaccharides are established prebiotic food ingredients used to promote the development of bifidobacteria and lactobacilli in the large intestine in order to reduce growth of pathogenic microorganisms [104].

Monosaccharides are applied as low-calorie sweeteners, bulking agents, antioxidants, glycosidase inhibitors, and nucleoside analogs. Production of monosaccharides can be catalyzed by isomerases. A recombinant mannose-6-phosphate isomerase from *G. thermodenitrificans* isomerizes aldose substrates possessing hydroxyl groups oriented in the same direction at the C2 and C3 positions such as the D- and L-forms of ribose, lyxose, talose, mannose, and allose. The enzyme was shown to catalyze the conversion of D-lyxose to D-xylulose with 38% yield after 3 h, and conversion of L-ribose to L-ribulose with 29% yield [105].

Keratin is the insoluble fibrous hard-to-degrade protein of feathers, wool, hair, including other epidermal appendages. It accounts for nearly 90–95% of feather weight, which constitutes up to 10% of the total chicken weight. The increased amounts of keratin by-products may represent a pollution problem. Keratinolytic proteinases or keratinases are known to utilize insoluble substrates such as fibrin, keratin, elastin, and collagen, and soluble substrates such as sodium caseinate, albumin, and gelatin. These enzymes have potential outlets in biomedicine, pharmaceutics, cosmetics, and waste bioconversion. A keratinolytic proteinase from *G. stearothermophilus* AD-11 shows optimal activity at pH 9 and 60°C and degrades keratin from wool > collagen > sodium caseinate > gelatin > and bovine serum albumin in descending order with production of high-value small peptides suitable for industrial applications [106].

Hydroxamic acids are weak organic acids. They are key pharmacophores in chemotherapeutic formulas possessing a wide spectrum of activities as growth factors, food additives, tumor inhibitors, antimicrobial, antituberculosis and antileukemic agents. Acetohydroxamic acid is a potent and irreversible inhibitor of bacterial and plant urease used in adjunctive therapy of chronic urinary infections. It also selectively shows anti-HIV activity and inhibits arachidonate 5-lipoxygenase, demonstrating potential use in the treatment of asthma. Intracellular amidase of *G. pallidus* BTP-5x MTCC 9225 showed complete conversion of acetamide to acetohydroxamic acid in 1 h at 50°C. At 65°C, the rate of reverse reaction was found to be higher. The acetamide bioconversion rate for the strain was 90–95% and 51 g powder containing 40% acetohydroxamic acid was recovered after lyophilization [107].

Chiral α-hydroxy acids are used in the production of pharmaceuticals and other fine chemicals. L-Lactate dehydrogenase from the thermophilic organism *G. stearothermophilus* may be employed for the industrial synthesis of chiral α-hydroxy acids. It is also possible to engineer the enzyme with enhanced activity toward the selected α-keto acids, besides natural substrates [108].

Recombinant alcohol dehydrogenase from *G. stearothermophilus* takes part in the biocatalytic synthesis of ω-oxo lauric acid methyl ester (OLAMe), a key intermediate in bio-based polyamide 12 production from the corresponding long-chain alcohol. The enzyme provides for up to 23% conversion of the substrate to OLAMe after 30 min. No overoxidation to the dodecanoic diacid monomethyl ester was detected. Thus, the engineered alcohol dehydrogenase is a promising biocatalyst for industrial polymer production [109].

Amidases are among the most widely used amide-hydrolyzing enzymes. They find use as catalysts in the treatment of industrial effluents containing toxic amides, in organic synthesis, and as therapeutic agents. The amidase of *G. pallidus* RAPc8 exhibited high thermal stability at 50 and 60°C, with half-lives over 5 h at both temperatures. At 70 and 80°C, the half-life values were 43 and 10 min, respectively. The enzyme catalyzed the hydrolysis of low molecular weight aliphatic amides, with D-selectivity toward lactamide. Acyl transfer reactions were demonstrated with acetamide, propionamide, isobutyramide, and acrylamide as substrates and hydroxylamine as the acyl acceptor. This amidase shows potential for application as a biocatalyst for D-selective amide hydrolysis yielding enantiomerically pure carboxylic acids and for the production of novel amides by acyl transfer [110]. Acrylamide is extensively used as a monomer in the synthesis of polyacrylamides, in dye composition, in gels for electrophoresis, in contact lenses, in food wraps, and in construction grouts. However, it displays many toxic properties such as neurotoxicity, genotoxicity, and carcinogenicity and can cause water and soil pollution. Acrylamidase from *G. thermoglucosidasius* AUT-01 is able to transform acrylamide to acrylic acid with pH and temperature optima of 6.2 and 70°C [111].

L-Nucleosides have been widely used as nucleoside-analog drugs in the treatment of severe viral diseases because they have more potent biological activities and lower toxicities than the corresponding D-nucleosides. L-Ribose is a potential starting material for the synthesis of many antiviral drugs, such as L-nucleoside derivatives. A triple-site variant of mannose-6-phosphate isomerase from *G. thermodenitrificans* demonstrated 3.1 and 7.1 times higher specific activity and catalytic efficiency, respectively, for L-ribulose isomerization compared to the wild-type enzyme at pH 7.0 and 70°C. The triple-site variant produced 213 g/l L-ribose from 300 g/l L-ribulose by 60 min, which exceeded 4.5 times the level of the wild-type enzyme. The specific activity, catalytic efficiency, and productivity of the variant were approximately two-fold higher than those of the *Thermus thermophilus* R142N isomerase, which exhibited the highest values previously reported [112]. Recombinant thermostable enzymes purine nucleoside phosphorylase II and pyrimidine nucleoside phosphorylase from *G. stearothermophilus* B-2194 retained high activity after 20 reuses in nucleoside transglycosylation reactions at 70–75°C with yields of the target products as high as 96%. These enzymes are suitable for the production of pharmacologically important natural and modified nucleosides [113].

Phytic acid is the most important storage form of inositol and phosphate in plants constituting approximately 5% of the dry weight of seeds in legumes and grain cereals. Phytic acid can form complexes with proteins and ions such as magnesium, calcium, zinc, and iron and act as antinutritional factor. Addition of phytases to animal feed can be an effective strategy to decrease phosphorus contamination and increase the bioavailability of phosphorus and essential minerals to animals. After 4 h incubation, hydrolysis capacity of chitosan- and Ca-alginate immobilized phytases of *Geobacillus* sp. TF-16 for soy milk phytate was calculated as 24 and 33%, respectively. The chitosan- and Ca-alginate immobilized enzymes conserved their original activity after 8 and 6 cycles of reuse, respectively [114].

Bacteria modulate their population density via the regulatory mechanism called quorum sensing. The latter takes part in bioluminescence, antibiosis, biofilm development, and control of expression of virulence genes. The Gram-negative bacteria use N-acylhomoserine lactone (AHL) as the quorum sensing signal. *G. caldoxylosilyticus* YS-8 was found to produce AHL lactonase degrading various AHLs. This enzyme can be further used in medical applications [115].

Some enzymes of *Geobacillus* species can be engaged in biotechnologies. GeoICI from *Geobacillus* sp., a member of atypical IIS restriction endonucleases, recognizing/cleaving 5′-GCAGC($N_{8/12}$)-3′ DNA sequences is highly active at elevated temperatures, up to 73°C and over a very wide salt concentration range, and hence can be applied in DNA manipulations [116]. β-Galactosidases are also objects for molecular applications in thermophiles and under anaerobic conditions. The enzyme from *G. stearothermophilus* functions both as a marker, when it cleaves thermostable dye, 3,4-cyclohexenoesculetin β-d-galactopyranoside (S-gal) to black product, and as a reporter enabling quantitative measurement by a simple colorimetric assay [117].

Aldehyde dehydrogenases are a group of diverse enzymes catalyzing the oxidation of aldehydes to carboxylic acids, using NAD^+ or $NADP^+$ as the coenzyme. The enzyme from *G. thermodenitrificans* NG80-2 demonstrated a broad substrate range including both aliphatic and aromatic aldehydes. It is expected to play a role in the degradation of alkanes and aromatic hydrocarbons present in crude oil. The aldehyde dehydrogenase activity was detected in the temperature range from 40 to 70°C, and in the pH range from 6.0 to 8.8. The optimum temperature was determined to be 60°C and the optimum pH 8.0. The enzyme was inactivated after incubation at 80°C [118].

Purification protocols for enzymes include several steps. The initial step is the release of the enzyme from the cell material, if the protein is not secreted by the organism into the surrounding solution. This procedure requires either a mechanical or chemical lysis of the cells. Techniques vary from gentle methods such as osmotic shock, detergent lysis, or enzymatic digestion to more vigorous methods

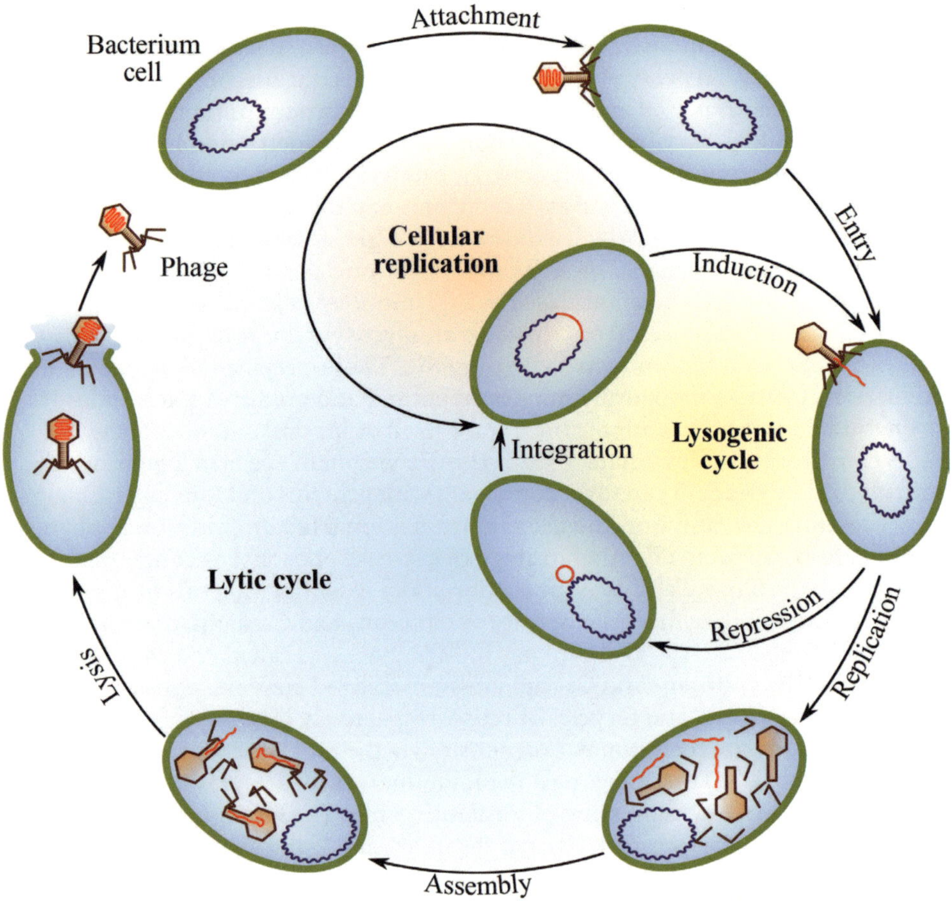

Figure 4.
Bacteriophage lytic and lysogenic cycles.

such as homogenization in a blender, grinding with an abrasive substance or ultrasonication. The selected procedure must not damage the target enzyme, and therefore, the conditions must be optimized for each cell type and target enzyme in terms of the pH of the extraction buffer, the temperature, and the concentration of certain components of the buffer such as detergents, salts, or reducing agents. The next step after clarification is the concentration of the enzyme preparation. Dialysis can be performed prior to this procedure in order to remove salts from the cell extract. The most common concentration procedures are ammonium sulfate precipitation, ultrafiltration, and ion exchange chromatography [119]. Purification of *Geobacillus* proteins follows the same steps. Cell-free supernatant of exo-α-1,4-glucosidase-producing strain *G. thermodenitrificans* F84a was removed from the medium by centrifugation, fractioned with solid ammonium sulfate, dialyzed overnight and centrifuged to remove insoluble residues. Then, the sample was concentrated 6.5-fold by centrifugation, suspended in Tris-HCl buffer (pH 8.0) and applied to cation-exchange chromatography [86]. The culture broth with amylase-producing *Geobacillus*sp. IIPTN was centrifuged; then supernatant was collected and filtered followed by ion-exchange chromatography [84]. Cells of amidase-producing *G. pallidus* RAPc8 were harvested by centrifugation and resuspended in an appropriate amount of potassium phosphate buffer. Cell lysis was achieved by freezing the cells at -20°C overnight, thawing at room temperature and sonication for 6–10 cycles. After subsequent centrifugation, the soluble fraction was heat-treated. Then, precipitated proteins were removed by centrifugation. Further purification was performed by gel exclusion chromatography [110].

Besides bacteria themselves, their bacteriophages are also potential sources of enzymes. Viruses constitute a major component of the biosphere, playing a significant role in nutrient and energy turnover of carbon, nitrogen, and phosphorus, and producing important impact on the evolution of their hosts. Thermophilic viruses are worse studied compared to the mesophilic viruses, and the majority of them infect archaeal genera. Generally, the life cycle of bacteriophage includes several programmed steps, such as phage adsorption to host cell surface, injection of phage genomic DNA into bacterial cell, metabolic transition from host to phage, phage genome replication, phage morphogenesis, phage package, and lysis of the host (**Figure 4**). Bacteriophage-host protein interactions in high-temperature environment remain poorly understood. Nevertheless, studies are carried out, like the discovery that the host's aspartate aminotransferase, chaperone GroEL, and viral capsid protein VP371 of bacteriophage GVE2 (host *Geobacillus* sp. E263) formed a linearly interacted complex for protection of the virus reproduction in high-temperature environment [120]. Because of adaptation to extreme conditions, such bacteriophages can be used as sources of ligases, polymerases, and nucleases for biotechnological processes [121].

6. Bioenergy technologies: the potential of biological processes

The growing demand for energy to keep up with the industrial spurt and the rampant urbanization has created a huge shortfall, urging to resort to alternative energy options. The global bioenergy market is anticipated to reach $246.52 billion by 2024 from $158.39 billion in 2015. During the forecast period of 2016–2024, the global market is expected to rise at a compound annual growth rate of 4.9% [122]. Bioenergy is renewable resource derived from organic matter (biomass), i.e., all materials of biological origin not embedded in geological formations (fossilized). Biomass can be used in its original form as fuel, or be refined to different kinds of solid, gaseous, or liquid biofuels. The supply of biomass can be provided from forestry, agriculture, industrial, and municipal waste [123].

Bioenergy is divided into three broad categories: solid biomass (e.g., wood, harvesting residues), liquid biofuels (e.g., bioethanol, biodiesel), and gaseous biofuels (e.g., biogas). Bioenergy accounted for 14% of the global energy consumption in 2012 with roughly 2.6 billion people dependent on traditional biomass for power supply (**Figure 5**). USA and Brazil lead the world in production and consumption of liquid biofuels for transport accounting for almost 80% of the market. The production of all biofuels in the Americas increased from about 16 billion liters in 2000 to 79 billion liters in 2012. The use of biomass for electricity is prominent in Europe and North America—predominantly produced from forestry products and residues. The Europe and American continent contribute more than 70% of overall consumption of biomass for electricity. In 2013, 462 TWh of electricity was produced globally from biomass. The major use of biomass is household heating in rural and developing countries [123].

Bioethanol is a biodegradable, renewable energy resource which is produced from biomass through sugar fermentation and chemical process. It forms an attractive alternative to conventional fuel sources owing to its high octane value and lower greenhouse gas emissions. Bioethanol can be used as a motor fuel, fuel for power generation, feedstock for chemical industry, substrate for fuel cells and cogeneration systems, in cosmetic technology, and in manufacturing processes due to its clean burning and ready availability [124]. The global bioethanol market is projected to reach USD 68.95 billion by 2022, at a compound annual growth rate of 5.3% between 2017 and 2022 [125].

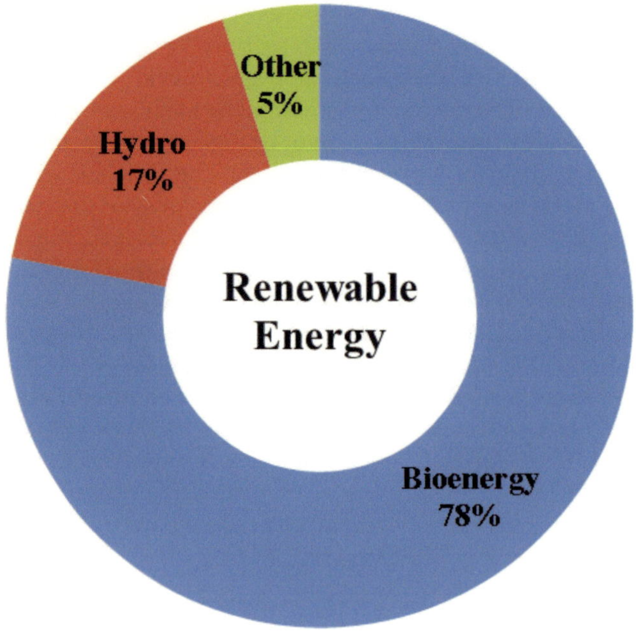

Figure 5.
Global renewable energy consumption.

Ethanol produced from renewable sources by fermentation is the most promising biofuel and the starting material for various chemicals. Substrates for ethanol production can be classified into three main groups: (1) those containing considerable amounts of readily fermentable sugars (sugarcane, sugar beet, and sweet sorghum), (2) starches and fructosans (corn, potatoes, rice, wheat, and agave), and (3) cellulosics (stover, grasses, corn cobs, wood, and sugarcane bagasse) [126]. Ethanol-producing microorganisms such as *Saccharomyces cerevisiae* and *Zymomonas mobilis* lack amylases and cellulases and are unable to directly convert starch and cellulosics into ethanol. Traditionally, the starch is hydrolyzed enzymatically into fermentable sugars via liquefaction and saccharification processes prior to ethanol fermentation [82].

Biodiesel is a mixture of long-chain monoalkylic esters of fatty acids obtained from renewable resources, to be used in diesel engines, alone or blended with diesel oil. The global biodiesel market size was estimated as USD 28.04 billion in 2016 and is likely to reach USD 54.8 billion by 2025. The market is expected to expand at 7.3% compound annual growth rate owing to increasing demand for biodiesel as automobile fuel with eco-friendly characteristics. The automotive fuel segment accounts for over 75% of the market [127]. The raw materials for biodiesel production are vegetable oils, animal fats, and short-chain alcohols. Biodiesel is produced by transesterification reaction. This chemical reaction converts vegetable oil or animal fat into a mixture of esters of the fatty acids that make up oil or fat. Biodiesel is derived by purification of the mixture of fatty acid methyl esters. Transesterification can be basic, acidic, or enzymatic. In the latter case, lipases are used [128].

Global isobutanol market demand was estimated at 552.4 kilo tons in 2014. It is used as a raw material for the production of various chemicals, including cleaners and coating solvents, isobutyl esters, extractants for pharmaceuticals, textiles, polish additive, gasoline admixture, agricultural products, and biofuel. The global isobutanol market is expected to reach USD 1.18 billion by 2022. Synthetic isobutanol was the largest product segment accounting for 58.1% of total market volume in 2014. Bio-based isobutanol is estimated to witness the highest growth of 7.0%

from 2015 to 2022 [129]. Isobutanol can be a better biofuel than ethanol due to its higher energy capacity and lower hygroscopicity. Furthermore, the branched-chain structure of isobutanol gives a higher octane number than the isomeric *n*-butanol. Bacteria have been shown to produce this compound, but isobutanol demonstrated toxicity to cells, forcing to use more resistant microorganisms [130].

Biogas, a mixture of methane and carbon dioxide, is produced from the methanogenic decomposition of organic waste under anaerobic conditions. Between 2009 and 2015, the number of biogas plants in Europe increased significantly from around 6000 to nearly 17,000. Total European Union biogas primary energy production in 2014 was estimated at 14.9 Mtoe, up 6.6% from the previous year. About 57 TWh of EU electricity was produced from biogas in 2014, up 9% from 2013. However, biogas share is only 1.9% of the total electricity generation. USA has been slower in launching biogas plants, with around 2200 in operation, of which the majority are run at wastewater treatment facilities [131].

Geobacillus species have been shown to produce or stimulate synthesis of some biofuels. Nevertheless, investigations of *Geobacillus* role in these processes started recently.

7. *Geobacillus* bacteria in bioenergy technologies and applications

The demand for energy production is at an all-time high level. Fossil materials, like coal and oil, are important fuels; however, they are considered as non-renewable resources raising serious environmental concerns. Bioenergy is the alternative source of fuels. Multiple approaches are currently explored for the use of microorganisms in the production of biofuel (alcohols, hydrogen, biodiesel, and biogas) from various starting materials. *Geobacillus* species able to synthesize a wide range of enzymes and resistant to high temperatures, allowing to minimize the risk of contamination and reduce energy consumption for product separation and fermenter cooling, appear excellent agents for bioenergy production.

Ethanol can be produced through fermentation by various strains of mesophilic bacteria, yeasts, and fungi. Thermophiles are able to utilize a wide range of sugars at high temperature converting them into ethanol. Most organisms in this class do not naturally carry out homoethanol fermentation and do not naturally exhibit high product tolerance, demanding additional metabolic engineering [132]. Thermophilic bacterium *G. thermoglucosidasius* M10EXG fermented a range of C5 (e.g., xylose) and C6 sugars (e.g., glucose) and was tolerant to high ethanol concentrations (10%, v/v). The carbon flux during microaerobic growth was directed to ethanol, L-lactate (>99% optical purity), acetate, and formate. Under fully anaerobic conditions, the strain was involved in a mixed acid fermentation process, achieving a maximum ethanol yield of 0.38 ± 0.07 mol/mol glucose [133]. Strains of *G. thermoglucosidasius* have been engineered to divert their fermentative carbon flux from a mixed acid pathway to that where ethanol becomes the major product. The triple mutant TM242 generated ethanol from glucose at yields higher than 90% of the theoretical value 0.51 g/g with productivity 2.85 g/l/h. In addition, it was particularly rapid in the metabolism of cellobiose (productivity 3.2 g/l/h) [134]. Bacterial enzymes can be used in ethanol production. About 20 and 30% raw pearl millet starch exposure to the above-mentioned hyperthermostable and Ca^{2+}-independent α-amylase of *G. thermoleovorans* NP54 for 3 h resulted in the sugar yields of 68 and 55.8%, respectively. Upon subsequent treatment with amylopullulanase of *G. thermoleovorans* NP33 for 4 h, 85 and 80% starch saccharification rates were attained from 20 and 30% raw starch, respectively. Saccharification was further enhanced to 98 and 92.4%, respectively, when the hydrolysate was treated with glucoamylase of *Thermomucor indicae-sedaticae* for 12 h. Following fermentation of reducing sugars in the hydrolysates, ethanol production levels by *S. cerevisiae* were 35.40 and 28.0 g/l [82].

As mentioned above, biodiesel is produced via transesterification reaction, and in some cases, lipases are used. Lipase from *G. thermodenitrificans* AV-5 showed molecular weight 50 kDa, temperature and pH optima, 65°C and pH 9.0, respectively, and was able to efficiently convert waste cooking oil and coconut oil to biodiesel with yields 76 and 45.5%, respectively [135]. The enzymatic reaction system enables to utilize low-grade and low-cost feedstock with high free fatty acid or water content. It reduces the amount of alkali wastewater that requires treatment and promotes easy recovery of the main product and the by-product glycerol. In addition, the biocatalyst is biodegradable and when immobilized can be reused for many production cycles and can lower the operation cost of the process. However, the major drawbacks of the enzymatic system are relatively high biocatalyst cost and its limited stability in the presence of high methanol concentrations. In order to achieve complete conversion of one oil molar equivalent to fatty acid methyl ester, three molar equivalents of methanol are necessary. Nevertheless, in reaction systems with more than 1.5 methanol equivalents, the methanol is not completely dissolved and its droplets stay in the mixture leading to enzyme unfolding and inactivation. The triple mutant lipase of *G. stearothermophilus* T6 showed a half-life value of 324 min in the presence of 70% methanol, which reflects 87-fold enhanced stability as compared to the wild type. This variant also exhibited an improved biodiesel yield from waste chicken oil when compared to commercial Lipolase 100L® and Novozyme® CALB [136]. The recombinant lipase from *G. stearothermophilus* G3 immobilized on the aminated silica gel can be used as a biocatalyst for the preparation of fatty acid methyl esters from vegetable oils. The optimum reaction parameters allowed to produce fatty acid methyl esters with 40–43% yield within 96 h [137].

Isobutanol biosynthesis shares intermediates with the valine biosynthesis pathway, which exists in most microorganisms, including *Geobacillus*. Engineered *G. thermoglucosidasius* variant produced 3.3 g/l of isobutanol from glucose and 0.6 g/l of isobutanol from cellobiose by 48 h at 50°C, demonstrating stable isobutanol generation in recombinant bacteria at an elevated temperature [138]. The other *G. thermoglucosidasius* culture was able to produce isobutanol in amounts around 50 mg/l via the conversion of isobutyryl-CoA to isobutyraldehyde by aldehyde dehydrogenase and from isobutyraldehyde to isobutanol by alcohol dehydrogenase. It was observed that supplementing the growth medium with an intermediate of the valine biosynthesis pathway, 2-ketoisovalerate, resulted in increased isobutanol titers [139].

Acetoin is widely used in food processing, flavoring, cosmetic formulation and chemical synthesis, while its reduced form, 2,3-butanediol, compares favorably with ethanol (29,100 J/g) and methanol (22,100 J/g) in heating value (27,200 J/g), so that 2,3-butanediol can be used as a liquid fuel or fuel additive. Its dehydration yields the industrial solvent methyl ethyl ketone, which can be hydrogenated to high octane isomers, suitable for high-quality aviation fuels. Acetoin and 2,3-butanediol are currently fermented below 40°C using mesophilic strains, but the processes often suffer from bacterial contamination. The application of thermophilic strains is able to meet the challenge. *Geobacillus* strain XT15 generated 7.7 g/l of acetoin and 14.5 g/l of 2,3-butanediol when incubated with corn steep liquor as a nitrogen source at 55°C. Acetoin, 2,3-butanediol, and their derivatives accounted for about 96% of total volatile products, while organic acids and other metabolites were minor by-products [140].

Due to its high organic content, sewage sludge is used as a substrate for anaerobic digestion to recover biogas. Anaerobic digestion of organic waste material is an effective technology for both waste disposal and energy generation. The addition of sludge with aerobic thermophilic (AT) bacteria (closely related to *G. thermodenitrificans*) to methanogenic sludge enhanced the production of biogas. The optimum added volume and the pretreatment temperature of the AT sludge for optimum biogas production were 5% and 65°C. The AT sludge inoculated with the AT seed sludge (mass of sludge containing populations of microorganisms) improved biogas production by 2.2 times. The addition of bacterial culture reduced volatile solids by 21%, which was higher than 12.6% achieved with the sewage sludge addition. The bacteria enhanced biogas production more than AT seed sludge [141]. The increased biogas production from anaerobic digestion of sewage sludge can be caused by the protease activity [142].

Microbial fuel cells (MFCs), another power-generating system, utilize bacteria acting as living catalysts to convert organic substrates into electricity. MFCs lack viability in most applications and demand optimization. Recently, it has been shown that *Geobacillus* species can be used for energy generation. However, *Geobacillus* strain could not produce current in the absence of an exogenous electron shuttle [143].

8. Conclusion

The *Geobacillus* species are represented by Gram-positive, aerobic or facultatively anaerobic, spore-forming, rod-shaped cells. The bacteria able to survive and grow at high temperatures can be found everywhere. At present, *Geobacillus* genus comprises about 20 species, but they are quite heterogeneous and require reclassification.

Owing to the ability of geobacilli to utilize and to produce a wide range of substances, coupled with resistance to high temperatures, they are considered as promising agents for many biological processes. *Geobacillus* genus is a source of a vast array of thermostable enzymes: amylases, lipases, pectinases, β-galactosidases, endonucleases, etc. Applications of bacteria and their enzymes range from food industry and medicine to molecular biology and bioremediation. New investigations concerning *Geobacillus* characteristics and spheres of use are carried out every year.

The growing demand for energy urges researchers and manufacturers to resort to alternative options. Bioenergy is the attractive source of fuels. It accounted for 14% of the global energy consumption in 2012. Multiple approaches are currently being probed to use microorganisms in the production of various fuels from diverse materials. *Geobacillus* genus demonstrated ability to produce or stimulate synthesis of some biofuels. Generation of biofuels coupled with resistance to high temperatures allows to enlist *Geobacillus* bacteria as potential candidates for bioenergy projects.

Acknowledgements

The work and chapter were funded through the FP7—Seventh Framework Programme of the European Community for Research and Technological Development (The ener2i project—Energy to Innovation—Reinforcing cooperation with ENP countries on bridging the gap between energy research and energy innovation. Reference of the Grant Agreement no: 609532).

Author details

Galina Novik[1*], Victoria Savich[1] and Olga Meerovskaya[2]

1 National Academy of Sciences of Belarus, Institute of Microbiology, Minsk, Belarus

2 Belarusian Institute of System Analysis and Information Support for Scientific and Technical Sphere, Minsk, Belarus

*Address all correspondence to: galina_novik@mbio.bas-net.by

References

[1] Logan NA, Vos PD, Dinsdale A. *Geobacillus*. In: Whitman WB, editor. Bergey's Manual of Systematics of Archaea and Bacteria. Hoboken, New Jersey: John Wiley & Sons; 2015. pp. 1-26. DOI: 10.1002/9781118960608.gbm00533

[2] Zeigler DR. The *Geobacillus* paradox: Why is a thermophilic bacterial genus so prevalent on a mesophilic planet? Microbiology. 2014;**160**(Pt 1):1-11. DOI: 10.1099/mic.0.071696-0

[3] Setlow P. Spores of *Bacillus subtilis*: Their resistance to and killing by radiation, heat and chemicals. Journal of Applied Microbiology. 2006;**101**(3):514-525. DOI: 10.1111/j.1365-2672.2005.02736.x

[4] Shapton DA, Hindes WR. The standardisation of a spore count technique. Chemistry and Industry. 1963;**41**(6):230-234

[5] Ren XY, Cai HS, Lin XJ, Qiu HD, Chen JC. Optimum medium composition for *Geobacillus stearothermophilus* CHB1 fermentation. Fujian Journal of Agricultural Sciences. 2007;**2007**(1):54-57

[6] Lee YH, Brown MRW, Cheung HY. Defined minimal media for the growth of prototrophic and auxotrophic strains of *Bacillus stearothermophilus*. Journal of Applied Bacteriology. 1982;**53**(2):179-187. DOI: 10.1111/j.1365-2672.1982.tb04675.x

[7] Ahmad S, Scopes RK, Rees GN, Patel BKC. *Saccharococcus caldoxylosilyticus* sp. nov., an obligately thermophilic, xylose-utilizing, endospore-forming bacterium. International Journal of Systematic and Evolutionary Microbiology. 2000;**50**(Pt 2):517-523. DOI: 10.1099/00207713-50-2-517

[8] Fortina MG, Mora D, Schumann P, Parini C, Manachini PL, Stackebrandt E. Re-classification of *Saccharococcus caldoxylosilyticus* as *Geobacillus caldoxylosilyticus* (Ah-mad et al. 2000) comb. nov. International Journal of Systematic and Evolutionary Micro-biology. 2001;**51**(Pt 6):2063-2071. DOI: 10.1099/00207713-51-6-2063

[9] Ibrahim MAC, Ahmad WA. Growth optimization of a thermophilic strain *Geobacillus caldoxylosilyticus* UTM6 isolated from Selayang hot spring. eProceedings Chemistry. 2017;**2**(1):119-123

[10] Adkins JP, Cornell LA, Tanner RS. Microbial composition of carbonate petroleum reservoir fluids. Geomicrobiology Journal. 1992;**10**(2):87-97. DOI: 10.1080/01490459209377909

[11] Nazina TN, Lebedeva EV, Poltaraus AB, Tourova TP, Grigoryan AA, Sokolova DS, Lysenko AM, Osipov GA. *Geobacillus gargensis* sp. nov., a novel thermophile from a hot spring, and the reclassification of *Bacillus vulcani* as *Geobacillus vulcani* comb. nov. International Journal of Systematic and Evolutionary Microbiology. 2004;**54**(Pt 6):2019-2024. DOI: 10.1099/ijs.0.02932-0

[12] Cuebas M, Sannino D, Bini E. Isolation and characterization of arsenic resistant *Geobacillus kaustophilus* strain from geothermal soils. Journal of Basic Microbiology. 2011;**51**(4):364-371. DOI: 10.1002/jobm.201000314

[13] Merkel GJ, Underwood WH, Perry JJ. Isolation of thermophilic bacteria capable of growth solely in long-chain hydrocarbons. FEMS Microbiology Letters. 1978;**3**(2):81-83

[14] Zarilla KA, Perry JJ. *Bacillus thermoleovorans*, sp. nov., a species of obligately thermophilic hydrocarbon utilizing endospore-forming bacteria.

Systematic and Applied Microbiology. 1987;**9**(3):258-264. DOI: 10.1016/S0723-2020(87)80031-0

[15] Pavlostathis SG, Marchant R, Banat IM, Ternan NG, McMullan G. High growth rate and substrate exhaustion results in rapid cell death and lysis in the thermophilic bacterium *Geobacillus thermoleovorans*. Biotechnology and Bioengineering. 2006;**95**(1):84-95. DOI: 10.1002/bit.20962

[16] Somerton B, Palmer J, Brooks J, Smolinski E, Lindsay D, Flint S. Influence of cations on growth of thermophilic *Geobacillus* spp. and *Anoxybacillus flavithermus* in planktonic culture. Applied and Environmental Microbiology. 2012;**78**(7):2477-2481. DOI: 10.1128/AEM.06951-11

[17] Nazina TN, Tourova TP, Poltaraus AB, Novikova EV, Grigoryan AA, Ivanova AE, Lysenko AM, Petrunyaka VV, Osipov GA, Belyaev SS, Ivanov MV. Taxonomic study of aerobic thermophilic bacilli: Descriptions of *Geobacillus subterraneus* gen. nov., sp. nov. and *Geobacillus uzenensis* sp. nov. from petroleum reservoirs and transfer of *Bacillus stearothermophilus*, *Bacillus thermocatenulatus*, *Bacillus thermoleovorans*, *Bacillus kaustophilus*, *Bacillus thermodenitrificans* to *Geobacillus* as the new combinations G. *stearothermophilus*, G. th. International Journal of Systematic and Evolutionary Microbiology. 2001;**51**(Pt 2):433-446. DOI: 10.1099/00207713-51-2-433

[18] Nazina TN, Sokolova DS, Grigoryan AA, Shestakova NM, Mikhailova EM, Poltaraus AB, Tourova TP, Lysenko AM, Osipov GA, Belyaev SS. *Geobacillus jurassicus* sp. nov., a new thermophilic bacterium isolated from a high-temperature petroleum reservoir, and the validation of the *Geobacillus* species. Systematic and Applied Microbiology. 2005;**28**(1):43-53. DOI: 10.1016/j.syapm.2004.09.001

[19] Golovacheva RS, Loginova LG, Salikhov TA, Kolesnikov AA, Zaĭtseva GN. New species of thermophilic bacilli—*Bacillus thermocatenulatus* nov. sp. Mikrobiologiia. 1975;**44**(2):265-268

[20] Souza AND, Martins MLL. Isolation, properties and kinetics of growth of a thermophilic *Bacillus*. Brazilian Journal of Microbiology. 2001;**32**(4):271-275. DOI: 10.1590/S1517-83822001000400003

[21] Dinsdale AE, Halket G, Coorevits A, Van Landschoot A, Busse HJ, De Vos P, Logan NA. Emended descriptions of *Geobacillus thermoleovorans* and *Geobacillus thermocatenulatus*. International Journal of Systematic and Evolutionary Microbiology. 2011;**61**(Pt 8):1802-1810. DOI: 10.1099/ijs.0.025445-0

[22] Mora D, Fortina MG, Nicastro G, Parini C, Manachini PL. Genotypic characterization of thermophilic bacilli: A study on new soil isolates and several reference strains. Research in Microbiology. 1998;**149**(10):711-722. DOI: 10.1016/S0923-2508(99)80018-7

[23] Baek DH, Lee YJ, Sin HS, Oh DK. A new thermophile strain of *Geobacillus thermodenitrificans* having L-arabinose isomerase activity for tagatose production. Journal of Microbiology and Biotechnology. 2004;**14**(2):312-316

[24] Ezeji TC, Wolf A, Bahl H. Isolation, characterization, and identification of *Geobacillus thermodenitrificans* HRO10, an alpha-amylase and alpha-glucosidase producing thermophile. Canadian Journal of Microbiology. 2005;**51**(8):685-693. DOI: 10.1139/w05-054

[25] Suzuki Y, Kishigami T, Abe S. Production of extracellular alpha-glucosidase by a thermophilic *Bacillus* species. Applied and Environmental Microbiology. 1976;**31**(6):807-812

[26] Suzuki Y, Kishigami T, Inoue K, Mizoguchi Y, Eto N, Takagi M, Abe S. *Bacillus thermoglucosidasius* sp. nov., a new species of obligately thermophilic bacilli. Systematic and Applied Microbiology. 1983;**4**(4):487-495. DOI: 10.1016/S0723-2020(83)80006-X

[27] Javed M, Baghaei-Yazdi N, Qin W, Amartey S. An improved agar medium for growth of *Geobacillus thermoglucosidasius* strains. Journal of Microbiological Methods. 2017;**132**:116-118. DOI: 10.1016/j.mimet.2016.11.014

[28] Kuisiene N, Raugalas J, Chitavichius D. *Geobacillus lituanicus* sp. nov. International Journal of Systematic and Evolutionary Microbiology. 2004;**54**(Pt 6):1991-1995. DOI: 10.1099/ijs.0.02976-0

[29] Sung MH, Kim H, Bae JW, Rhee SK, Jeon CO, Kim K, Kim JJ, Hong SP, Lee SG, Yoon JH, Park YH, Baek DH. *Geobacillus toebii* sp. nov., a novel thermophilic bacterium isolated from hay compost. International Journal of Systematic and Evolutionary Microbiology. 2002;**52**(Pt 6):2251-2255. DOI: 10.1099/00207713-52-6-2251

[30] Poli A, Romano I, Caliendo G, Nicolaus G, Orlando P, Falco A, Lama L, Gambacorta A, Nicolaus B. *Geobacillus toebii* subsp. *decanicus* subsp. nov., a hydrocarbon-degrading, heavy metal resistant bacterium from hot compost. The Journal of General and Applied Microbiology. 2006;**52**(4):223-234. DOI: 10.2323/jgam.52.223

[31] Caccamo D, Gugliandolo C, Stackebrandt E, Maugeri TL. *Bacillus vulcani* sp. nov., a novel thermophilic species isolated from a shallow marine hydrothermal vent. International Journal of Systematic and Evolutionary Microbiology. 2000;**50**(Pt 6):2009-2012. DOI: 10.1099/00207713-50-6-2009

[32] Poli A, Laezza G, Gul-Guven R, Orlando P, Nicolaus B. *Geobacillus galactosidasius* sp. nov., a new thermophilic galactosidase-producing bacterium isolated from compost. Systematic and Applied Microbiology. 2011;**34**(6):419-423. DOI: 10.1016/j.syapm.2011.03.009

[33] Bryanskaya AV, Rozanov AS, Slynko NM, Shekhovtsov SV, Peltek SE. *Geobacillus icigianus* sp. nov., a thermophilic bacterium isolated from a hot spring. International Journal of Systematic and Evolutionary Microbiology. 2015;**65**(Pt 3):864-869. DOI: 10.1099/ijs.0.000029

[34] Donk PJ. A highly resistant thermophilic organism. Journal of Bacteriology. 1920;**5**(4):373-374

[35] Ash C, Farrow JAE, Wallbanks S, Collins MD. Phylogenetic heterogeneity of the genus *Bacillus* revealed by comparative analysis of small-subunit-ribosomal RNA sequences. Letters in Applied Microbiology. 1991;**13**(4):202-206. DOI: 10.1111/j.1472-765X.1991.tb00608.x

[36] List of Prokaryotic Names with Standing in Nomenclature: LPSN [Internet]. 2018. Available from: http://www.bacterio.net/geobacillus.html [Accessed: Feb 10, 2018]

[37] Zeigler DR. Application of a *recN* sequence similarity analysis to the identification of species within the bacterial genus *Geobacillus*. International Journal of Systematic and Evolutionary Microbiology. 2005;**55**(Pt 3):1171-1179. DOI: 10.1099/ijs.0.63452-0

[38] Studholme DJ. Some (bacilli) like it hot: Genomics of *Geobacillus* species. Microbial Biotechnology. 2015;**8**(1):40-48. DOI: 10.1111/1751-7915.12161

[39] Aliyu H, Lebre P, Blom J, Cowan D, De Maayer P. Phylogenomic

re-assessment of the thermophilic genus *Geobacillus*. Systematic and Applied Microbiology. 2016;**39**(8):527-533. DOI: 10.1016/j.syapm.2016.09.004

[40] Tourova TP, Korshunova AV, Mikhailova EM, Sokolova DS, Poltaraus AB, Nazina TN. Application of *gyrB* and *parE* sequence similarity analyses for differentiation of species within the genus *Geobacillus*. Microbiology. 2010;**79**(3):356-369. DOI: 10.1134/S0026261710030124

[41] Weng FY, Chiou CS, Lin PH, Yang SS. Application of *recA* and *rpoB* sequence analysis on phylogeny and molecular identification of *Geobacillus* species. Journal of Applied Microbiology. 2009;**107**(2):452-464. DOI: 10.1111/j.1365-2672.2009.04235.x

[42] Kuisiene N, Raugalas J, Chitavichius D. Phylogenetic, inter, and intraspecific sequence analysis of *spo0A* gene of the genus *Geobacillus*. Current Microbiology. 2009;**58**(6):547-553. DOI: 10.1007/s00284-009-9366-2

[43] Kuisiene N, Raugalas J, Stuknyte M, Chitavichius D. Identification of the genus *Geobacillus* using genus-specific primers, based on the 16S-23S rRNA gene internal transcribed spacer. FEMS Microbiology Letters. 2007;**277**(2):165-172. DOI: 10.1111/j.1574-6968.2007.00954.x

[44] Krilaviciute A, Kuisiene N. Characterization and evaluation of tandem repeats for the identification of *Geobacillus*. Central European Journal of Biology. 2013;**8**(6):549-560. DOI: 10.2478/s11535-013-0168-3

[45] Adiguzel A, Ozkan H, Baris O, Inan K, Gulluce M, Sahin F. Identification and characterization of thermophilic bacteria isolated from hot springs in Turkey. Journal of Microbiological Methods. 2009;**79**(3):321-328. DOI: 10.1016/j.mimet.2009.09.026

[46] Seale RB, Dhakal R, Chauhan K, Craven HM, Deeth HC, Pillidge CJ, Powell IB, Turner MS. Genotyping of present day and historical *Geobacillus* spp. isolates from milk powders using high resolution melt analysis (HRMA) of multiple variable number tandem repeat (VNTR) loci. Applied and Environmental Microbiology. 2012;**78**(19):7090-7097. DOI: 10.1128/AEM.01817-12

[47] Kuisiene N, Raugalas J, Čitavičius D. Identification of *Geobacillus stearothermophilus* by restriction the amplified 16S rDNA with *AluI*. Biologija. 2007;**53**(4):62-66. DOI: 10.6001/biologija.v53i4.767

[48] Zhou JF, Gao PK, Dai XH, Cui XY, Tian HM, Xie JJ, Li GQ, Ma T. Heavy hydrocarbon degradation of crude oil by a novel thermophilic *Geobacillus stearothermophilus* strain A-2. International Biodeterioration and Biodegradation. 2018;**126**:224-230. DOI: 10.1016/j.ibiod.2016.09.031

[49] Sun Y, Ning Z, Yang F, Li X. Characteristics of newly isolated *Geobacillus* sp. ZY-10 degrading hydrocarbons in crude oil. Polish Journal of Microbiology. 2015;**64**(3):253-263

[50] Wenjie X, Li Y, Ping W, Jianlong X, Hanping D. Characterization of a thermophilic and halotolerant *Geobacillus pallidus* H9 and its application in microbial enhanced oil recovery (MEOR). Annals of Microbiology. 2012;**62**(4):1779-1789. DOI: 10.1007/s13213-012-0436-5

[51] Banat IM, Marchant R. Geobacillus activities in soil and oil contamination remediation. In: Logan NA, De Vos P, editors. Endospore-Forming Soil Bacteria. Berlin, Heidelberg: Springer; 2011. pp. 259-270. DOI: 10.1007/978-3-642-19577-8_13

[52] Pranckute R, Kaunietis A, Kananaviciute R, Lebedeva J, Kuisiene

N, Saleikiene J, Citavicius D. Differences of antibacterial activity spectra and properties of bacteriocins, produced by *Geobacillus* sp. bacteria isolated from different environments. The Journal of Microbiology, Biotechnology and Food Sciences. 2015;**5**(2):155-161. DOI: 10.15414/jmbfs.2015.5.2.155-161

[53] Garg N, Tang W, Goto Y, Nair SK, van der Donk WA. Lantibiotics from *Geobacillus thermodenitrificans*. Proceedings of the National Academy of Sciences of the United States of America. 2012;**109**(14):5241-5246. DOI: 10.1073/pnas.1116815109

[54] Arivalagan K, Ravichandran S, Rangasamy K, Karthikeyan E. Nanomaterials and its potential applications. International Journal of ChemTech Research. 2011;**3**(2):534-538

[55] Salata O. Applications of nanoparticles in biology and medicine. Journal of Nanobio-technology. 2004;**2**:3. DOI: 10.1186/1477-3155-2-3

[56] Fayaz AM, Girilal M, Rahman M, Venkatesan R, Kalaichelvan PT. Biosynthesis of silver and gold nanoparticles using thermophilic bacterium *Geobacillus stearothermophilus*. Process Biochemistry. 2011;**46**(10):1958-1962. DOI: 10.1016/j.procbio.2011.07.003

[57] Ghasemi SM, Dormanesh B, Hosseini-Abari AS, Aliasghari A, Farahnejad Z. Comparative characterization of silver nanoparticles synthesized by spore extract of *Bacillus subtilis* and *Geobacillus stearothermophilus*. Nanomedicine Journal. 2018;**5**(1):46-51. DOI: 10.22038/nmj.2018.05.007

[58] Nara K, Iwata K, Matsui T, Shigeno T, Omori T. Functional analysis of the thermophilic denitrifying bacterium *Geobacillus* sp. strain TDN01 in continuous culture. The Journal of General and Applied Microbiology. 2009;**55**(2):87-92. DOI: 10.2323/jgam.55.87

[59] Mahendrakar K, Venkategowda PM, Rao SM, Mutkule DP. Glyphosate surfactant herbicide poisoning and management. Indian Society of Critical Care Medicine. 2014;**18**(5):328-330. DOI: 10.4103/0972-5229.132508

[60] Obojska A, Ternan NG, Lejczak B, Kafarski P, McMullan G. Organophosphonate utilization by the thermophile *Geobacillus caldoxylosilyticus* T20. Applied and Environmental Microbiology. 2002;**68**(4):2081-2084. DOI: 10.1128/AEM.68.4.2081-2084.2002

[61] Evangelista-Barreto NS, Albuquerque CD, Vieira RHS, Campos-Takaki GM. Cometabolic decolorization of the reactive azo dye orange II by *Geobacillus stearothermophilus* UCP 986. Textile Research Journal. 2009;**79**(14):1266-1273. DOI: 10.1177/0040517508087858

[62] Tomita K, Ikeda N, Ueno A. Isolation and characterization of a thermophilic bacterium, *Geobacillus thermocatenulatus*, degrading nylon 12 and nylon 66. Biotechnology Letters. 2003;**25**(20):1743-1746. DOI: 10.1023/A:1026091711130

[63] Özdemir S, Gul-Guven R, Kilinc E, Dogru M, Erdogan S. Preconcentration of cadmium and nickel using the bioadsorbent *Geobacillus thermoleovorans* subsp. *stromboliensis* immobilized on Amberlite XAD-4. Microchimica Acta. 2010;**169**(1-2):79-85. DOI: 10.1007/s00604-010-0300-x

[64] Chatterjee SK, Bhattacharjee I, Chandra G. Biosorption of heavy metals from industrial waste water by *Geobacillus thermodenitrificans*. Journal of Hazardous Materials. 2010;**175**(1-3):117-125. DOI: 10.1016/j.jhazmat.2009.09.136

[65] Özdemir S, Kılınç E, Poli A, Nicolaus B. Biosorption of heavy metals (Cd^{2+}, Cu^{2+}, Co^{2+}, and Mn^{2+}) by thermophilic bacteria, *Geobacillus thermantarcticus* and *Anoxybacillus amylolyticus*: Equilibrium and kinetic studies. Bioremediation Journal. 2013;**17**(2):86-96. DOI: 10.1080/10889868.2012.751961

[66] Arena A, Gugliandolo C, Stassi G, Pavone B, Iannello D, Bisignano G, Maugeri TL. An exopolysaccharide produced by *Geobacillus thermodenitrificans* strain B3-72: Antiviral activity on immunocompetent cells. Immunology Letters. 2009;**123**(2):132-137. DOI: 10.1016/j.imlet.2009.03.001

[67] Jara AM, Andrade RF, Campos-Takaki GM. Physicochemical characterization of tensio-active produced by *Geobacillus stearothermophilus* isolated from petroleum-contaminated soil. Colloids and Surfaces. B, Biointerfaces. 2013;**101**:315-318. DOI: 10.1016/j.colsurfb.2012.05.021

[68] Smerilli M, Neureiter M, Wurz S, Haas C, Frühauf S, Fuchs W. Direct fermentation of potato starch and potato residues to lactic acid by *Geobacillus stearothermophilus* under non-sterile conditions. Journal of Chemical Technology and Biotechnology. 2015;**90**(4):648-657. DOI: 10.1002/jctb.4627

[69] Mohd Azam RMA, Mohd Saman H, Kamaruddin K, Hussain N. Enhancement of thermophilic (*Geobacillus stearothermophilus*) cement–sand mortar properties. In: Yacob N, Mohamed M, Megat Hanafiah M, editors. Regional Conference on Science, Technology and Social Sciences (RCSTSS 2014). Singapore: Springer; 2016. pp. 79-92. DOI: 10.1007/978-981-10-0534-3_8

[70] Kim JJ, Masui R, Kuramitsu S, Seo JH, Kim K, Sung MH. Characterization of growth-supporting factors produced by *Geobacillus toebii* for the commensal thermophile *Symbiobacterium toebii*. Journal of Microbiology and Biotechnology. 2008;**18**(3):490-496

[71] Lü Y, Li N, Yuan X, Hua B, Wang J, Ishii M, Igarashi Y, Cui Z. Enhancing the cellulose-degrading activity of cellulolytic bacteria CTL-6 (*Clostridium thermocellum*) by co-culture with non-cellulolytic bacteria W2-10 (*Geobacillus* sp.). Applied Biochemistry and Biotechnology. 2013;**171**(7):1578-1588. DOI: 10.1007/s12010-013-0431-8

[72] Kader AEA, Aly MS, Esawy MA. Inoculation of *Geobacillus caldoxylosilyticus* IRD into maize (*Zea mays* L.) to sustain tolerance against high salt stress. Journal of American Science. 2011;7(1):71-79

[73] Park TJ, Lee SJ, Park JP, Yang MH, Choi JH, Lee SY. Characterization of a bacterial self-assembly surface layer protein and its application as an electrical nanobiosensor. Journal of Nanoscience and Nanotechnology. 2011;**11**(1):402-407. DOI: 10.1166/jnn.2011.3264

[74] Kumar A. The Probiotics Market: Ingredients, Supplements, Foods. 2018. Available from:https://www.bccresearch.com/market-research/food-and-beverage/probiotics-market-ingredients-supplements-foods-report-fod035e.html [Accessed: Sep 20, 2017]

[75] Mahdhi A, Hmila Z, Behi A, Bakhrouf A. Preliminary characterization of the probiotic properties of *Candida famata* and *Geobacillus thermoleovorans*. Iranian Journal of Microbiology. 2011;**3**(3):129-134

[76] Mau T, Hartmann V, Burmeister J, Langguth P, Häusler H. Development

of a sterilizing in-place application for a production machine using vaporized hydrogen peroxide. PDA Journal of Pharmaceutical Science and Technology. 2004;**58**(3):130-146

[77] Kawamura K, Sakuma A, Nakamura Y, Oguri T, Sato N, Kido N. Evaluation of bactericidal effects of low-temperature nitrogen gas plasma towards application to short-time sterilization. Microbiology and Immunology. 2012;**56**(7):431-440. DOI: 10.1111/j.1348-0421.2012.00457.x

[78] Szulc M, Schein S, Schaup J, Zimmermann S, Schein J. Suitability of thermal plasmas for large-area bacteria inactivation on temperature-sensitive surfaces—First results with *Geobacillus stearothermophilus* spores. Journal of Physics: Conference Series. 2017;**825**(1):012017. DOI: 10.1088/1742-6596/825/1/012017

[79] Blanchard K, Robic S, Matsumura I. Transformable facultative thermophile *Geobacillus stearothermophilus* NUB3621 as a host strain for metabolic engineering. Applied Micro-biology and Biotechnology. 2014;**98**(15):6715-6723. DOI: 10.1007/s00253-014-5746-z

[80] Jayani RS, Saxena S, Gupta R. Microbial pectinolytic enzymes: A review. Process Biochemistry. 2005;**40**(9):2931-2944. DOI: 10.1016/j.procbio.2005.03.026

[81] Çelik SY, Demir N, Demir Y, Adiguzel A, Gulluce M. Production of pectin lyase from *Geobacillus pallidus* p26, purification, characterization and fruit juice application. Acta Chimica Slovaca. 2014;**7**(1):57-63. DOI: 10.2478/acs-2014-0011

[82] Rao JUM, Satyanarayana T. Improving production of hyperthermostable and high maltose-forming α-amylase by an extreme thermophile *Geobacillus thermoleovorans* using response surface methodology and its applications. Bioresource Technology. 2007;**98**(2):345-352. DOI: 10.1016/j.biortech.2005.12.022

[83] Rao JUM, Satyanarayana T. Statistical optimization of a high maltose-forming, hyperthermostable and Ca^{2+}-independent alpha-amylase production by an extreme thermophile *Geobacillus thermoleovorans* using response surface methodology. Journal of Applied Microbiology. 2003;**95**(4):712-718. DOI: 10.1046/j.1365-2672.2003.02036.x

[84] Dheeran P, Kumar S, Jaiswal YK, Adhikari DK. Characterization of hyperthermostable alpha-amylase from *Geobacillus* sp. IIPTN. Applied Microbiology and Biotechnology. 2010;**86**(6):1857-1866. DOI: 10.1007/s00253-009-2430-9

[85] Khemakhem B, Ali MB, Aghajari N, Juy M, Haser R, Bejar S. Engineering of the alpha-amylase from *Geobacillus stearothermophilus* US100 for detergent incorporation. Biotechnology and Bioengineering. 2009;**102**(2):380-389. DOI: 10.1002/bit.22083

[86] Cihan AC, Ozcan B, Tekin N, Cokmus C. Characterization of a thermostable α-glucosidase from *Geobacillus thermodenitrificans* F84a. In: Mendez-Vilas A, editor. Current Research, Technology and Education Topics in Applied Microbiology and Microbial Biotechnology, Microbiology Books Series. Vol. 2. Badajoz, Spain: Formatex Research Center; 2010. pp. 945-955

[87] Hung VS, Hatada Y, Goda S, Lu J, Hidaka Y, Li Z, Akita M, Ohta Y, Watanabe K, Matsui H, Ito S, Horikoshi K. Alpha-Glucosidase from a strain of deep-sea *Geobacillus*: A potential enzyme for the biosynthesis of complex carbohydrates. Applied Microbiology

and Biotechnology. 2005;**68**(6):757-765. DOI: 10.1007/s00253-005-1977-3

[88] Ng IS, Li CW, Yeh YF, Chen PT, Chir JL, Ma CH, Yu SM, Ho TH, Tong CG. A novel endo-glucanase from the thermophilic bacterium *Geobacillus* sp. 70PC53 with high activity and stability over a broad range of temperatures. Extremophiles. 2009;**13**(3):425-435. DOI: 10.1007/s00792-009-0228-4

[89] Assareh R, Shahbani Zahiri H, Akbari Noghabi K, Aminzadeh S, Bakhshi Khaniki G. Characterization of the newly isolated *Geobacillus* sp. T1, the efficient cellulase-producer on untreated barley and wheat straws. Bioresource Technology. 2012;**120**:99-105. DOI: 10.1016/j.biortech.2012.06.027

[90] Tai SK, Lin HP, Kuo J, Liu JK. Isolation and characterization of a cellulolytic *Geobacillus thermoleovorans* T4 strain from sugar refinery wastewater. Extremophiles. 2004;**8**(5):345-349. DOI: 10.1007/s00792-004-0395-2

[91] Bhardwaj KK, Saun NK, Gupta R. Immobilization of lipase from *Geobacillus* sp. and its application in synthesis of methyl salicylate. Journal of Oleo Science. 2017;**66**(4):391-398. DOI: 10.5650/jos.ess16153

[92] Abol Fotouh DM, Bayoumi RA, Hassan MA. Production of thermoalkaliphilic lipase from *Geobacillus thermoleovorans* DA2 and application in leather industry. Enzyme Research. 2016;**2016**:1-9. DOI: 10.1155/2016/9034364

[93] Balan A, Ibrahim D, Rahim RA. Organic-solvent and surfactant tolerant thermostable lipase, isolated from a thermophilic bacterium, *Geobacillus thermodenitrificans* IBRL-nra. Advanced Studies in Biology. 2013;**5**(9):389-401. DOI: 10.12988/asb.2013.3624

[94] Nik Raikhan NH. Purification of a thermostable lipase *Geobacillus thermodenitrificans* nr68 (Lip. nr-68) with potential for enzymatic deinking. Journal of Life Sciences and Technologies. 2016;**4**(2):70-74. DOI: 10.18178/jolst.4.2.70-74

[95] Qin XL, Huang HH, Lan DM, Wang YH, Yang B. Typoselectivity of crude *Geobacillus* sp. T1 lipase fused with a cellulose-binding domain and its use in the synthesis of structured lipids. Journal of the American Oil Chemists' Society. 2014;**91**(1):55-62. DOI: 10.1007/s11746-013-2347-4

[96] Sun S, Lu F, Song H, Xu J, Jiang N, Zhang Y, Li L, Ming T, Zheng W. Cloning, expression and characterization of a thermostable pullulanase from newly isolated thermophilic *Geobacillus* sp. LM14-3. In: 4th International Conference on Biomedical Engineering and Informatics (BMEI), Shanghai, China. Vol. 3. October 15-17, 2011. pp. 1567-1570. DOI: 10.1109/BMEI.2011.6098577

[97] Canakcı S, Cevher Z, Inan K, Tokgoz M, Bahar F, Kacagan M, Sal FA, Belduz AO. Cloning, purification and characterization of an alkali-stable endoxylanase from thermophilic *Geobacillus* sp. 71. World Journal of Microbiology and Biotechnology. 2012;**28**(5):1981-1988. DOI: 10.1007/s11274-011-1000-3

[98] Sharma A, Adhikari S, Satyanarayana T. Alkali-thermostable and cellulase-free xylanase production by an extreme thermophile *Geobacillus thermoleovorans*. World Journal of Microbiology and Biotechnology. 2007;**23**(4):483-490. DOI: 10.1007/s11274-006-9250-1

[99] Cakmak U, Ertunga NS. Gene cloning, expression, immobilization and characterization of endo-xylanase from *Geobacillus* sp. TF16 and investigation

of its industrial applications. Journal of Molecular Catalysis B: Enzymatic. 2017;**133**(Suppl 1):S288-S298. DOI: 10.1016/j.molcatb.2017.01.016

[100] Jain I, Kumar V, Satyanarayana T. Applicability of recombinant β-xylosidase from the extremely thermophilic bacterium *Geobacillus thermodenitrificans* in synthesizing alkylxylosides. Bioresource Technology. 2014;**170**:462-469. DOI: 10.1016/j.biortech.2014.07.113

[101] Canakci S, Belduz AO, Saha BC, Yasar A, Ayaz FA, Yayli N. Purification and characterization of a highly thermostable alpha-L-arabinofuranosidase from *Geobacillus caldoxylolyticus* TK4. Applied Microbiology and Biotechnology. 2007;**75**(4):813-820. DOI: 10.1007/s00253-007-0884-1

[102] Derekova A, Atanassova M, Christova P, Tchorbanov B, Shosheva A, Mandeva R, Rodríguez-Alonso P, Garabal JI, Kambourova M. Physicochemical characteristics of a thermostable gellan lyase from *Geobacillus stearothermophilus* 98. Zeitschrift für Naturforschung. Section C. 2010;**65**(3-4):231-238. DOI: 10.1515/znc-2010-3-411

[103] Nath A, Mondal S, Chakraborty S, Bhattacharjee C, Chowdhury R. Production, purification, characterization, immobilization, and application of β-galactosidase: A review. Asia-Pacific Journal of Chemical Engineering. 2014;**9**(3):330-348. DOI: 10.1002/apj.1801

[104] Placier G, Watzlawick H, Rabiller C, Mattes R. Evolved beta-galactosidases from *Geobacillus stearothermophilus* with improved transgalactosylation yield for galacto-oligosaccharide production. Applied and Environmental Microbiology. 2009;**75**(19):6312-6321. DOI: 10.1128/AEM.00714-09

[105] Yeom SJ, Kim NH, Yoon RY, Kwon HJ, Park CS, Oh DK. Characterization of a mannose-6-phosphate isomerase from *Geobacillus thermodenitrificans* that converts monosaccharides. Biotechnology Letters. 2009;**31**(8):1273-1278. DOI: 10.1007/s10529-009-0003-8

[106] Gegeckas A, Gudiukaitė R, Debski J, Citavicius D. Keratinous waste decomposition and peptide production by keratinase from *Geobacillus stearothermophilus* AD-11. International Journal of Biological Macromolecules. 2015;**75**:158-165. DOI: 10.1016/j.ijbiomac.2015.01.031

[107] Sharma M, Sharma NN, Bhalla TC. Biotransformation of acetamide to acetohydroxamic acid at bench scale using acyl transferase activity of amidase of *Geobacillus pallidus*BTP-5x MTCC 9225. Indian Journal of Microbiology. 2012;**52**(1):76-82. DOI: 10.1007/s12088-011-0211-5

[108] Aslan AS, Birmingham WR, Karagüler NG, Turner NJ, Binay B. Semi-rational design of *Geobacillus stearothermophilus* L-lactate dehydrogenase to access various chiral α-hydroxy acids. Applied Biochemistry and Biotechnology. 2016;**179**(3):474-484. DOI: 10.1007/s12010-016-2007-x

[109] Kirmair L, Seiler DL, Skerra A. Stability engineering of the *Geobacillus stearothermophilus* alcohol dehydrogenase and application for the synthesis of a polyamide 12 precursor. Applied Microbiology and Biotechnology. 2015;**99**(24):10501-10513. DOI: 10.1007/s00253-015-6930-5

[110] Makhongela HS, Glowacka AE, Agarkar VB, Sewell BT, Weber B, Cameron RA, Cowan DA, Burton SG. A novel thermostable nitrilase superfamily amidase from *Geobacillus pallidus* showing acyl transfer activity. Applied Microbiology and Biotechnology. 2007;**75**(4):801-811. DOI: 10.1007/s00253-007-0883-2

[111] Cha M, Chambliss GH. Cloning and sequence analysis of the heat-stable acrylamidase from a newly isolated thermophilic bacterium, *Geobacillus thermoglucosidasius* AUT-01. Biodegradation. 2013;**24**(1):57-67. DOI: 10.1007/s10532-012-9557-6

[112] Lim YR, Yeom SJ, Oh DK. Production of L-ribose from L-ribulose by a triple-site variant of mannose-6-phosphate isomerase from *Geobacillus thermodenitrificans*. Applied and Environmental Microbiology. 2012;**78**(11):3880-3884. DOI: 10.1128/AEM.07012-11

[113] Taran SA, Verevkina KN, Feofanov SA, Miroshnikov AI. Enzymatic transglycosylation of natural and modified nucleosides by immobilized thermostable nucleoside phosphorylases from *Geobacillus stearothermophilus*. Russian Journal of Bioorganic Chemistry. 2009;**35**(6):739-745. DOI: 10.1134/S1068162009060107

[114] Şirin Y, Yildirim Akatin M, Colak A, Ertunga N. Dephytinization of food stuffs by the phytase of *Geobacillus* sp. TF-16 immobilized in chitosan and calcium-alginate. International Journal of Food Properties. 2017;**20**(12):2911-2922. DOI: 10.1080/10942912.2016.1261151

[115] Seo MJ, Lee BS, Pyun YR, Park H. Isolation and characterization of N-acylhomoserine lactonase from the thermophilic bacterium, *Geobacillus caldoxylosilyticus* YS-8. Bioscience, Biotechnology, and Biochemistry. 2011;**75**(9):1789-1795. DOI: 10.1271/bbb.110322

[116] Zebrowska J, Zolnierkiewicz O, Skowron MA, Zylicz-Stachula A, Jezewska-Frackowiak J, Skowron PM. A putative Type IIS restriction endonuclease GeoICI from *Geobacillus* sp.—A robust, thermostable alternative to mezophilic prototype BbvI. Journal of Biosciences. 2016;**41**(1):27-38. DOI: 10.1007/s12038-016-9595-z

[117] Jensen TØ, Pogrebnyakov I, Falkenberg KB, Redl S, Nielsen AT. Application of the thermostable β-galactosidase, BgaB, from *Geobacillus stearothermophilus* as a versatile reporter under anaerobic and aerobic conditions. AMB Express. 2017;**7**(1):169. DOI: 10.1186/s13568-017-0469-z

[118] Li X, Li Y, Wei D, Li P, Wang L, Feng L. Characterization of a broad-range aldehyde dehydrogenase involved in alkane degradation in *Geobacillus thermodenitrificans* NG80-2. Microbiological Research. 2010;**165**(8):706-712. DOI: 10.1016/j.micres.2010.01.006

[119] Dako E, Bernier AM, Dadie AT, Jankowski CK. The problems associated with enzyme purification. In: Ekinci D, editor. Chemical Biology. Rijeka, Croatia: InTech; 2012. pp. 19-40. DOI: 10.5772/33307

[120] Chen Y, Wei D, Wang Y, Zhang X. The role of interactions between bacterial chaperone, aspartate aminotransferase, and viral protein during virus infection in high temperature environment: The interactions between bacterium and virus proteins. BMC Microbiology. 2013;**13**:48. DOI: 10.1186/1471-2180-13-48

[121] Uldahl K, Peng X. Biology, biodiversity and application of thermophilic viruses. In: Satyanarayana T, Littlechild J, Kawarabayasi Y, editors. Thermophilic Microbes in Environmental and Industrial Biotechnology. Dordrecht: Springer; 2013. pp. 271-304. DOI: 10.1007/978-94-007-5899-5_10

[122] Bioenergy Market—Global Industry Analysis, Size, Share, Growth Trends and Forecast 2016-2024 [Internet]. 2016. Available from: https://www.transparencymarketresearch.com/global-bioenergy-market.html [Accessed: Feb 14, 2018]

[123] World Energy Council. Bioenergy: World Energy Resources 2016 [Internet]. 2016. Available from: https://www.worldenergy.org/publications/2016/world-energy-resources-2016/ [Accessed: Feb 14, 2018]

[124] Grand View Research. Bioethanol Market Analysis, Market Size, Application, Analysis, Regional Outlook, Competitive Strategies and Forecasts, 2015 to 2022 [Internet]. 2013. Available from: https://www.grandviewresearch.com/industry-analysis/bioethanol-market [Accessed: Feb 14, 2018]

[125] ReportsnReports. Bioethanol Market by Feedstock (Starch-Based, Sugar-Based, Cellulose-Based), End-Use Industry (Transportation, Pharmaceuticals, Cosmetics, Alcoholic Beverages), Blend (E5, E10, E15 to E70, E75 to E85), and Region—Global Forecast to 2022 [Internet]. 2017. Available from: http://www.reportsnreports.com/reports/1171868-bioethanol-market-by-feedstock-starch-based-sugar-based-cellulose-based-end-use-industry-transportation-pharmaceuticals-cosmetics-alcoholic-beverages-blend-e5-e10-e15-to-e70-e75-to-e85-and-region-st-to-2022.html [Accessed: Feb 14, 2018]

[126] Basso LC, Basso TO, Rocha SN. Ethanol production in Brazil: The industrial process and its impact on yeast fermentation. In: Bernardes MADS, editor. Biofuel Production—Recent Developments and Prospects. Rijeka, Croatia: InTech; 2011. pp. 85-100. DOI: 10.5772/17047

[127] Grand View Research, Inc. Biodiesel Market Analysis by Feedstock [Vegetable Oils (Canola Oil, Soybean Oil, Palm Oil, Corn Oil), Animal Fats (Poultry, Tallow, White Grease)], by Application (Fuel, Power Generation), and Segment Forecasts, 2014-2025 [Inter-net]. 2017. Available from: https://www.grandviewresearch.com/industry-analysis/biodiesel-market [Accessed: Feb 14, 2018]

[128] Romano SD, Sorichetti PA. Introduction to biodiesel production. In: Romano SD, Sorichetti PA, editors. Dielectric Spectroscopy in Biodiesel Production and Characterization, Green Energy and Technology. London: Springer; 2010. pp. 7-27. DOI: 10.1007/978-1-84996-519-4_2

[129] Grand View Research, Inc. Isobutanol Market Analysis by Product (Synthetic, Bio-Based), Application (Oil & Gas, Solvents & Coatings, Chemical Intermediates) and Segment Forecasts to 2022 [Internet]. 2016. Available from: https://www.grandviewresearch.com/industry-analysis/isobutanol-market [Accessed: Feb 14, 2018]

[130] Chen X, Nielsen KF, Borodina I, Kielland-Brandt MC, Karhumaa K. Increased isobutanol production in *Saccharomyces cerevisiae* by overexpression of genes in valine metabolism. Biotechnology for Biofuels. 2011;**4**:21. DOI: 10.1186/1754-6834-4-21

[131] Lambert M. Biogas: A Significant Contribution to Decarbonising Gas Markets? [Internet]. 2017. Available from: https://www.oxfordenergy.org/publications/biogas-significant-contribution-decarbonising-gas-markets [Accessed: Feb 14, 2018]

[132] Olson DG, Sparling R, Lynd LR. Ethanol production by engineered thermophiles. Current Opinion in Biotechnology. 2015;**33**:130-141. DOI: 10.1016/j.copbio.2015.02.006

[133] Tang YJ, Sapra R, Joyner D, Hazen TC, Myers S, Reichmuth D, Blanch H, Keasling JD. Analysis of metabolic pathways and fluxes in a newly discovered thermophilic and ethanol-tolerant *Geobacillus* strain. Biotechnology and Bioengineering.

2009;**102**(5):1377-1386. DOI: 10.1002/bit.22181

[134] Cripps RE, Eley K, Leak DJ, Rudd B, Taylor M, Todd M, Boakes S, Martin S, Atkinson T. Metabolic engineering of *Geobacillus thermoglucosidasius* for high yield ethanol production. Metabolic Engineering. 2009;**11**(6):398-408. DOI: 10.1016/j.ymben.2009.08.005

[135] Christopher LP, Zambare VP, Zambare A, Kumar H, Malek L. A thermo-alkaline lipase from a new thermophile *Geobacillus thermodenitrificans* AV-5 with potential application in biodiesel production. Journal of Chemical Technology and Biotechnology. 2015;**90**(11):2007-2016. DOI: 10.1002/jctb.4678

[136] Dror A, Kanteev M, Kagan I, Gihaz S, Shahar A, Fishman A. Structural insights into methanol-stable variants of lipase T6 from *Geobacillus stearothermophilus*. Applied Microbiology and Biotechnology. 2015;**99**(22):9449-9461. DOI: 10.1007/s00253-015-6700-4

[137] Samoylova YV, Piligaev AV, Sorokina KN, Rozanov AS, Peltek SE, Novikov AA, Almyasheva NR, Parmon VN. Application of the immobilized bacterial recombinant lipase from *Geobacillus stearothermophilus* G3 for the production of fatty acid methyl esters. Catalysis in Industry. 2016;**8**(2):187-193. DOI: 10.1134/S2070050416020082

[138] Lin PP, Rabe KS, Takasumi JL, Kadisch M, Arnold FH, Liao JC. Isobutanol production at elevated temperatures in thermophilic *Geobacillus thermoglucosidasius*. Metabolic Engineering. 2014;**24**:1-8. DOI: 10.1016/j.ymben.2014.03.006

[139] Martínez-Klimova E. Synthetic biology approaches to the metabolic engineering of *Geobacillus thermoglucosidans* for isobutanol production [thesis]. London: Imperial College; 2014

[140] Xiao Z, Wang X, Huang Y, Huo F, Zhu X, Xi L, Lu JR. Thermophilic fermentation of acetoin and 2,3-butanediol by a novel *Geobacillus* strain. Biotechnology for Biofuels. 2012;**5**(1):88. DOI: 10.1186/1754-6834-5-88

[141] Miah MS, Tada C, Yang Y, Sawayama S. Aerobic thermophilic bacteria enhance biogas production. Journal of Material Cycles and Waste Management. 2005;**7**(1):48-54. DOI: 10.1007/s10163-004-0125-y

[142] Miah MS, Tada C, Sawayama S. Enhancement of biogas production from sewage sludge with the addition of *Geobacillus* sp. strain AT1 culture. Japanese Journal of Water Treatment Biology. 2004;**40**(3):97-104. DOI: 10.2521/jswtb.40.97

[143] Wrighton KC, Agbo P, Warnecke F, Weber KA, Brodie EL, DeSantis TZ, Hugenholtz P, Andersen GL, Coates JD. A novel ecological role of the *Firmicutes* identified in thermophilic microbial fuel cells. The ISME Journal. 2008;**2**(11):1146-1156. DOI: 10.1038/ismej.2008.48

Chapter 2

Growing and Handling of Bacterial Cultures within a Shared Core Facility for Integrated Structural Biology Program

Lisa R. Warner, Olga Mass, Nancy Donnelly Lenn, Briana R. Grantham and Julia Thom Oxford

Abstract

We have established and optimized standard operating procedures for growing and handling bacterial cultures in a shared core laboratory to support Integrative Structural Biology. The Integrative Structural Biology effort within the Biomolecular Research Center allows researchers to generate new knowledge about protein and RNA structure and function. We aim to understand how biomolecules assemble into stable structures and how structural dynamics impacts their function. Here we describe specific procedures for growing and handling bacterial cultures for overexpression and isolation of recombinant proteins, $^{15}N/^{13}C$ uniform labeling of recombinant proteins, protein isolation and purification, and analysis of protein solubility that are ideal for implementation in a shared research core laboratory that serves a multitude of diverse customers and research laboratories.

Keywords: protein expression, recombinant protein, isotype enrichment, core facility, protocols

1. Introduction

Shared research core facilities can provide support to campus-wide investigators by providing research infrastructure for the production and purification of recombinant proteins for a variety of research applications. We have designed a research support structure for investigators pursuing research in structural and functional studies that require high yields of pure proteins, particularly suited for structural studies including biomolecular nuclear magnetic resonance (NMR) and small angle X-ray scattering.

The *Escherichia coli* (*E. coli*) expression platform is commonly used for recombinant expression of proteins. The *E. coli* system has several advantages over yeast, insect cells, or mammalian cell expression systems: *E. coli* are relatively easy to handle, the doubling time is short, media are low-cost and there are abundantly established methods for protein expression [1–4]. The *E. coli* expression platform is also well-suited for stable isotope labeling of proteins for biological NMR studies [5–9]. Structural studies of proteins demand large quantities of high purity protein. Meeting these requirements can be challenging, however, advancements in high-throughput technologies for recombinant expression of proteins have greatly advanced in the last decade or more, in large part due to efforts from large structural genomics and structural proteomics centers [1, 4, 10–12]. The lessons learned and technologies developed from these centers can allow for rapid assessment of different expression strategies, which can be transferred and scaled down to smaller-scale centers and academic labs [3, 4, 13].

In addition to a demand for large quantities of highly pure protein, structural studies also often demand high solubility and stability of the protein in solution. To address this need, a high-throughput fluorescence-based thermal-shift assay, also known as differential scanning fluorimetry (DSF), has been implemented at the large structural genomics and structural proteomics centers [14]. DSF was originally developed as a high-throughput drug discovery assay to screen for small molecules that bind to and stabilize target proteins [15–17]. The DSF screen has been further adapted to optimize buffer conditions by varying the pH, buffer components, detergents, reducing agents and small molecules to screen for conditions that increase the stability and conformational homogeneity of a protein [14, 17–20], which is key in obtaining high-quality structural data.

We have established and optimized standard operating procedures for growing and handling bacterial cultures in a shared core laboratory to support Integrative Structural Biology and have used these in our own research [21–29]. The Integrative Structural Biology effort within the Biomolecular Research Center, a shared core facility, allows researchers at Boise State University and collaborating institutions to generate new knowledge about protein and RNA structure and function. We aim to understand how biomolecules assemble into stable structures and how structural dynamics can impact their function. Here we describe specific procedures for growing and handling bacterial cultures for overexpression and isolation of recombinant proteins, $^{15}N/^{13}C$ uniform labeling of recombinant proteins, protein isolation and purification, and analysis of protein solubility that are ideal for implementation in a shared research core laboratory that serves a multitude of diverse customers

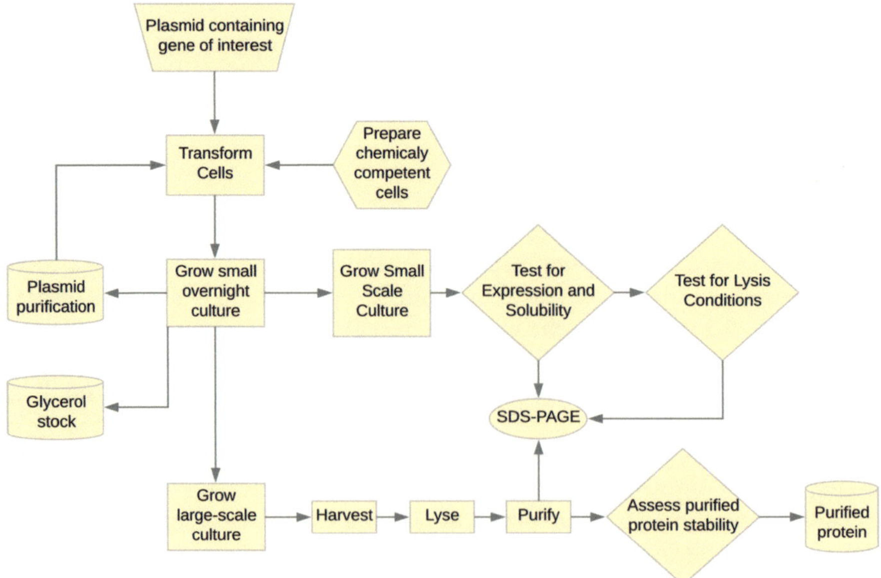

Figure 1.
A protein expression and purification workflow from plasmid to stable purified protein.

and research laboratories. Here we outline a general workflow of essential steps in protein expression and purification that includes plasmid amplification, mini-expression screening, optimized larger-scale protein production, protein isolation and purification, and characterization of optimized experimental solution buffer conditions (**Figure 1**).

2. Materials

All reagents listed in this chapter are commonly available from commercial vendors. A chemical hygiene plan including storage, shelf life, and safety of all chemicals should be in place at the institution.

2.1 Preparation of chemically competent cells

1. Ice.

2. Lysogeny broth (LB) medium: 10 g/L tryptone, 5 g/L yeast extract, 10 g/L NaCl. Sterilize by autoclaving and store at room temperature.

3. Super optimal broth (SOB, a.k.a. Hanahan's Broth) medium for DH5α cells: 20 g/L tryptone, 5 g/L yeast extract, 0.5 g/L NaCl, 0.186 g/L KCl. Adjust the pH to 7.0 with NaOH. Sterilize by autoclaving and store at room temperature.

4. Culture tubes and flasks.

5. Incubator/shaker.

6. Centrifuge tubes.

7. Serological pipettes.

8. Repeating pipettor.

9. Dimethyl sulfoxide (DMSO).

10. Competent Cell (CC) buffer: 10 mM HEPES (4-(2-hydroxyethyl)-1-piperazineethanesulfonic acid), 15 mM $CaCl_2$, 55 mM $MnCl_2$, 250 mM KCl, pH 6.7. Dissolve all components except $MnCl_2$ and adjust the pH to 6.7 with KOH. Then add the $MnCl_2$ and filter sterilize the solution over a 0.22 μm filter.

2.2 Transformation of cells for expression of desired plasmid

1. Ice.

2. LB or super optimal broth with catabolite repression (SOC).

3. Culture tubes and flasks.

4. Incubator/shaker.

5. Centrifuge tubes.

6. Serological pipettes.

7. Competent cells.

8. LB-agar plates containing the appropriate antibiotic.

9. Plasmid DNA.

10. Heat block set.

2.3 Calculating efficiency of competent cells

1. Transformed colonies on LB-agar plate (see Section 3.3).

2.4 Inoculating overnight cultures

1. LB.

2. 15 mL conical tube.

3. Sterile inoculating loop.

4. Appropriate antibiotics.

5. Shaker/incubator.

6. Sterile aluminum foil or culture tube cap.

2.5 Glycerol stocks

1. 50% glycerol solution (autoclaved).

    a. Make the 50% glycerol solution by diluting 100% glycerol into water.

2. Screwtop cryogenic vials.

3. Liquid nitrogen.

2.6 DNA plasmid purification

1. Resuspension buffer: 50 mM Tris-HCl, pH 8.0; 10 mM ethylenediaminetetraacetic acid (EDTA), 20 μg/mL RNase A.

2. Lysis buffer: 200 mM NaOH, 1% w/v sodium dodecyl sulfate (SDS).

3. Precipitation buffer: 3 M potassium acetate, 2 M glacial acetic acid, 4°C.

4. Wash buffer: 70% ethanol.

5. 95% (or 100%) ethanol.

6. TE buffer: 10 mM Tris-HCl, pH 8.0; 0.1 mM EDTA.

2.7 Testing for protein expression and solubility in *E. coli*

1. LB.

2. Appropriate antibiotics.

3. Incubator/shaker.

4. Microcentrifuge tubes.

5. Centrifuge.

6. Isopropyl β-D-1-thiogalactopyranoside (IPTG).

2.8 Lysing cells

1. Induced cells suspended in lysis buffer with a protease inhibitor cocktail, 0.1 mg/mL DNase I, 1 mg/mL lysozyme and 0.1 mg/mL 4-(2-Aminoethyl) benzenesulfonyl fluoride (AEBSF).

2. Sonication buffer.

3. Ice-saltwater bath.

4. Probe sonicator equipped with ½ inch tip.

2.9 Gel electrophoresis, protein quantification

1. Electrophoresis system.

2. 4–12% Bis-Tris mini gel.

3. Sample loading buffer: 10% glycerol, 0.14 M Tris Base, 0.1 M Tris-HCl, 2% lithium dodecyl sulfate (LDS), 0.5 mM EDTA, 0.02% Blue G250; 0.006% phenol red, 1.25% 2-mercaptoethanol, pH 8.5.

4. Running buffer: 50 mM 2-(N-morpholino)ethanesulfonic acid (MES), 50 mM Tris base, 0.1% SDS, 1 mM EDTA, pH 7.2.

5. Coomassie Blue stain.

6. Protein molecular weight marker.

2.10 Testing lysis conditions for solubility

1. Buffer A: 50 mM Tris pH 7.5, 100 mM NaCl, 5 mM EDTA, 1 mg/mL lysozyme.

2. Buffer B: 50 mM Tris pH 7.5, 2 M NaCl, 5 mM EDTA, 1 mg/mL lysozyme.

3. Buffer C: 50 mM Tris pH 7.5, 100 mM NaCl, 50% detergent, 1 mg/mL lysozyme.

2.11 Large-scale expression of recombinant proteins

1. LB.
2. IPTG.
3. Culture tubes and flasks.
4. Incubator/shaker.

2.12 Uniform $^{15}N/^{13}C$ labeling of recombinant proteins

1. LB.
2. IPTG.
3. Culture tubes and flasks.
4. Incubator/shaker.
5. 10X M9 medium: 340 mM Na_2HPO_4, 220 mM KH_2PO_4, 85.5 mM NaCl, pH 7.4.
6. 10X Ammonium chloride: 93.5 mM NH_4Cl.
7. 20% wt/vol glucose stock.
8. 100 mM $CaCl_2$.
9. 1.0 M $MgSO_4$.
10. 10 mg/mL thiamine.
11. 10 mg/mL biotin.[1]
12. Antibiotic for plasmid selection.

2.13 Protein purification using immobilized metal affinity chromotography (IMAC)

Immobilized metal affinity chromatography (IMAC) is a common method for affinity purification. A genetically encoded 6-histidine repeat affinity tag can be introduced to the carboxy or amino terminal end of the protein during cloning, which has high affinity for metal ions. The protocol given here is for affinity purification by immobilization of nickel ions with a chelator molecule, nitrilotriacetic acid (NTA) that is covalently bound to agarose; commonly known as Ni-NTA agarose. The following buffers are meant to represent a general starting point. Depending on the pI of your recombinant protein and the propensity to nonspecifically interact with the column material or resident *E. coli* proteins, modifications may need to be made. Additional purification may be necessary, especially when purifying proteins that bind to nucleic acids. A lithium wash may be added to the Ni-NTA purification

[1] The stock solution of 10 mg/mL is above the solubility limit of biotin, do not sterile filter this solution. Simply make the solution with previously sterilized water.

to remove nucleic acids. Ion exchange, heparin affinity, size exclusion chromatography are often added in addition to a nickel affinity purification step.

1. Lysis buffer: 0.1 M Tris-HCl, 0.1 M NaCl, pH 8.1.

2. Wash buffer: Lysis buffer plus 5–20 mM imidazole.

3. Elution buffer: Lysis buffer plus 100–300 mM imidazole.

4. Probe sonicator.

5. 1 mg/mL lysozyme.

6. Protease inhibitor cocktail.

7. AEBSF.

2.14 Differential scanning fluorimetry to assess protein stability

1. Low ionic strength buffer (e.g., 10 mM Tris-HCl).

2. qPCR machine with filter set that matches fluorescent dye and equipped with a ramp rate of minimum 1°C/min.

3. A fluorescent dye that will bind proteins.

4. 96-well polymerase chain reaction (PCR) microplate.

3. Methods

3.1 Preparation of chemically competent cells

1. Inoculate 5 mL of LB (or SOB if preparing DH5α cells) with 10 μL of appropriate *E. coli* cells[2] and grow overnight at 37°C and 250 rpm in a shaking incubator.

2. Use the overnight culture to inoculate 250 mL of LB (or SOB if preparing DH5α cells) and incubate at 30°C until the optical density at 600 nm (OD_{600}) is between 0.4–0.6.

3. Chill the culture for at least 10 min on ice. For steps 4–10, keep the cell suspension on ice.

4. Spin the cell suspension for 10 min at 6000× g.

5. Gently resuspend the pellet in 50 mL ice-cold CC buffer into 50-mL conical tubes. Resuspend with a 10-mL serological pipette and avoid introducing bubbles.

---

[2] Some cell lines have a resident plasmid, such as BL21(DE3) pLysS or pLysE cells and require addition of antibiotics for selection of cells containing those plasmids.

6. Incubate the cell suspension on ice for at least 10 min.

7. Spin for 10 min at 6000× g at 4°C.

8. Gently resuspend the pellet in 9.4 mL ice-cold CC buffer and add 0.7 mL DMSO.

9. Incubate the cell suspension on ice for at least 10 min.

10. Distribute the cell suspension in 50–200 μL aliquots in 1.5-mL microcentrifuge tubes.[3]

11. Flash freeze the cell suspension in liquid nitrogen and store the tubes at −80°C.

12. At −80°C the cells will be competent for at least 6 months.

3.2 Transformation of *E. coli* cells with plasmid DNA

1. Take competent cells out of −80°C and thaw on ice (approximately 20–30 min).

2. For each transformation, remove two LB-agar plates (containing the appropriate antibiotic) from storage at 4°C and warm to room temperature; optionally warm to 37°C in an incubator.

3. Mix 10–100 pg. DNA into 20–50 μL of competent cells in a 1.5 mL microcentrifuge tube.

4. Gently mix by flicking the bottom of the tube with your finger a few times.

5. Incubate the competent cell/DNA mixture on ice for 20–30 min.

6. Heat shock each tube at 42°C for 45–60 s.

7. Put the tubes back on ice for 2 min.

8. Add 1 mL of LB medium (without antibiotic) to the bacteria and grow at 37°C and 250 rpm in a shaking incubator for 45 min.

9. Plate 50 μL of the transformed cells onto one of the 10 cm LB-agar plate containing the appropriate antibiotic and the remaining 950 μL onto the second 10 cm LB-agar plate.

10. Incubate plates at 37°C overnight.

3.3 Calculating transformation efficiency of competent cells

1. Count the number of colony forming units (CFUs) on the LB-agar plate after transformation (see Section 3.2).

[3] A repeating pipettor or a multichannel pipettor speeds up the aliquoting process greatly. This will minimize the time that the competent cells are manipulated, thus increasing their competency. Expect competency of ca. 10^7–10^8 CFU/μg of plasmid DNA.

2. Calculate the transformation efficiency (TrEff) in CFUs/µg of DNA using Eq. (1).

$$TrEff = \frac{\#\ of\ CFUs \times Total\ vol.of\ Transformation\ (\mu L)}{\mu L\ of\ transformation\ on\ plate \times \mu g\ of\ DNA} \quad (1)$$

3.4 Inoculating cultures

1. Add 5–10 mL of liquid LB to a culture tube and add the appropriate antibiotic to at correct concentration. A good negative control is LB media plus antibiotic without any bacteria inoculated. You should see no growth in this culture after overnight incubation.

2. Using a sterile inoculating loop, select a single colony from your LB-agar plate for plasmid purifications and a swipe from 10 to 20 colonies for protein expression (Section 3.2).

3. Add the inoculating loop to the liquid LB with antibiotics and swirl.

4. Loosely cover the culture with sterile aluminum foil or a culture tube cap.

5. After incubation, check for growth, which is characterized by a cloudy haze in the media.

6. For overnight cultures, incubate bacterial culture at 30°C for 12–16 h in a shaking incubator.[4]

7. For long-term storage of the bacteria, you can proceed with Section 3.5.

3.5 Preparation of a glycerol stock

1. Follow Section 3.2 for transforming and plating *E. coli* cells.

2. Follow Section 3.4 for inoculating an overnight culture.

3. Add 500 µL of the overnight culture to 500 µL of 50% glycerol in a 2 mL screw top cryogenic vial[5] and gently mix.

4. Submerse the glycerol stock tube into liquid nitrogen and store at −80°C. The stock is now stable for years, as long as it is kept at −80°C. Subsequent freeze and thaw cycles reduce shelf life.

5. To recover bacteria from your glycerol stock: open the tube and use a sterile loop, toothpick, or pipette tip to scrape some of the frozen bacteria off of the top. Do not let the glycerol stock thaw. Streak the bacteria onto an LB-agar plate.

3.6 Plasmid DNA purification

1. Preheat the TE Buffer in the incubator at 37°C.

---

[4] For some applications (especially culturing cells in minimal defined media) cultures should never be overgrown; growing overnight cultures at a reduced temperature, 25–30°C, is suggested.

[5] Snap top tubes are not recommended.

2. Spin 5 mL of the overnight LB culture at 6000× g for 10 min at 4°C. Discard supernatant.

3. Add 250 μL Resuspension Buffer containing RNase A to the cell pellet and resuspend the pellet by pipetting until homogeneous.

4. Add 250 μL Lysis Buffer. Mix gently by inverting the capped tube until homogeneous. Do not vortex. Incubate the tube at room temperature for 5 min.

5. Add 350 μL Precipitation Buffer. Mix immediately by inverting the tube until homogeneous. Do not vortex. Centrifuge the lysate at 16,000× g for 10 min at 4°C.

6. Add 2–2.5 volumes 95% or 100% ethanol and 1/10 volume of 3 M Na-acetate (pH 4.8) to the supernatant. Invert the microcentrifuge tube to mix. Let stand for 2 min at room temperature.

7. Centrifuge solution at high speed (at least 16,000× g) for 15–30 min at 4°C. Discard supernatant.

8. Open and invert the tube on a paper towel to drain it out.

9. Wash pellet by adding 500 μL cold 70% ethanol.

10. Centrifuge solution at high speed (at least 16,000× g) for 5 min at room temp. Discard supernatant by pipetting it out of the tube.

11. Dry the pellet by inverting over paper towel for 5–20 min.

12. Resuspend dry DNA with TE.

13. Store plasmid DNA at 4°C (short term) or store the DNA in aliquots at −20°C (long term.)

3.7 Testing for soluble protein expression in *E. coli*

The following protocol is written for proteins expressed under the control of the lac, tac, or T7 promoters. The method as described is a generic protocol that can be expanded to test expression in different strains of *E. coli*, induction temperatures, concentrations of IPTG, or even in the presence of ligands or cofactors.

3.7.1 Protein expression

1. Transform plasmid into an *E. coli* expression strain following Section 3.2.

2. Inoculate a liquid LB culture following Section 3.4.

3. Grow cells for a few hours at 37°C, shaking at 250 rpm. Make sure the tubes are tilted.

4. Monitor the turbidity. Once the culture reaches an OD_{600} of 0.4–0.6 (takes ~2–4 h, depending on the sample), take out 2 mL of the culture. Measure the actual OD_{600}.

5. Transfer the equivalent of 1 mL of cells at OD_{600} = 0.8 in a 1.5-mL microcentrifuge tube.[6]

6. Collect cells by centrifugation at 16,000× g on a tabletop centrifuge for at least 1 min. Carefully remove all of the supernatant. This is the uninduced sample. Store the cells at −20°C.

7. Add IPTG to a final concentration of 1.0 mM to the remaining culture. Continue shaking at 250 rpm for 12–16 h at 18°C.

8. Measure the OD_{600}. Collect cells by centrifugation in two tubes containing the equivalent of 1 mL of cells at OD_{600} = 0.8 and remove the supernatant. These are the induced samples; one tube will be used to test for expression and the second for solubility. Store the cells at −20°C until ready to test for expression.

3.7.2 Testing for expression

1. Take the tube of uninduced and one tube of induced cells and resuspend each in 100 μL of 1X SDS polyacrylamide gel electrophoresis (SDS-PAGE) sample buffer.

2. Boil the samples for 10 min, then cool down to room temperature.

3. Centrifuge for 5 min at 16,000× g on a tabletop centrifuge at room temperature.

4. Take 10 μL of each sample from the top of tube taking care not to disturb the pellet.

5. Analyze the results using SDS-PAGE following Section 3.9 (**Figure 2**), with western blotting if necessary.

3.7.3 Testing for solubility

1. Resuspend the remaining induced cell pellet in 50 μL of lysis buffer containing protease inhibitors and 1 mg/mL of lysozyme.

2. Follow Section 3.8.1 for freeze-thawing to lyse the cells.

3. Spin down in a microcentrifuge at maximum speed for 10 min at 4°C.

4. Carefully transfer all of the supernatant into a new microcentrifuge tube. Add 50 μL of 2X SDS-PAGE buffer. This is the soluble fraction.

5. Resuspend the pellet in 100 μL of 1X SDS-PAGE buffer. This is the insoluble fraction.

6. Boil the samples for 10 min, then cool down to room temperature.

[6] In order to have equal loading on an SDS-PAGE gel, the same amount of cells need to be harvested for gel analysis. To harvest the same number of cells each time, calculate the volume in mL needed of your culture that would be the equivalent of 1 mL of OD600 = 0.8. E.g. X mL = 0.8/OD600 of your culture.

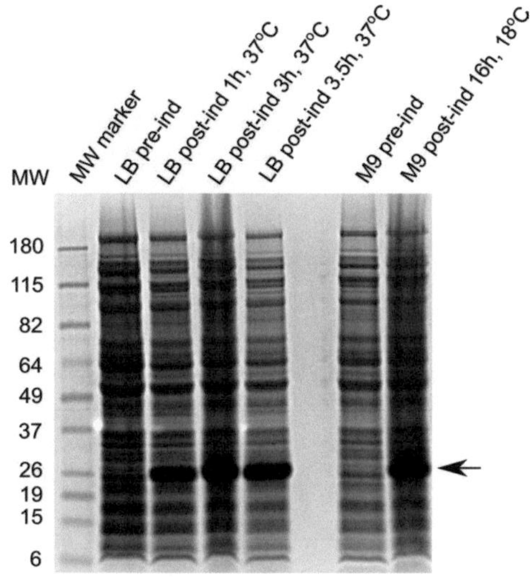

Figure 2.
SDS-PAGE gel of pre- and post-induction of an RNA binding protein (RBP) in both rich (LB) and minimal (M9) media. The arrow indicates the recombinant RBP.

7. Centrifuge for 5 min at 16,000× g at room temperature.

8. Analyze 15 μL of each sample using SDS-PAGE following Section 3.9.

3.8 Lysing cells

Traditionally cell lysis can be done with physical disruption or reagent-based methods. Freeze-thaw protocol works best for small volumes (less than 1 mL) in 1.5 mL microcentrifuge tubes. Sonication can be done with smaller volumes using a microtip.

3.8.1 Freeze-thaw

1. Freeze the samples to be lysed (typically 0.1–1.0 mL in a 1.5 mL microcentrifuge tube) in a − 80°C freezer, leave for 15 min.

2. Thaw immediately in a 42°C water bath. Vortex vigorously to mix well.

3. Repeat the two previous steps three more times (four freeze-thaw-vortex cycles in all).

4. Spin the tubes for 5 min at maximum speed in a microcentrifuge.

5. Separate the supernatant (contains soluble protein) from the pellet (contains insoluble protein) by pipetting out the supernatant to a clean tube.

3.8.2 Sonication

1. Prepare ice-saltwater bath by sprinkling salt over packed ice in a container.

2. Place a 50-mL conical tube containing the cell pellet suspended in lysis buffer securely in the ice-saltwater bath.

3. Insert clean tip of a sonicator in the sample without contacting sides or bottom of the tube.

4. Set the output power, cycle, and timer to the optimal settings (e.g., five short bursts of 15 s followed by intervals of 30 s for cooling).

5. Keep the suspension at all times on ice.

3.9 Gel electrophoresis

1. Add ~100 μg of protein to SDS sample buffer.

2. Heat the sample at 70°C for 10 min.

3. Load the entire volume of sample onto a 4–12% Bis-Tris mini gel.

4. Run the gel at 200 V for 35 min.

5. At the end of the electrophoresis, wash the gel in deionized water three times.

6. Stain the gel with Coomassie Blue stain for 1 h.

7. Wash the gel with deionized water extensively until the water is clear.

3.10 Testing lysis conditions for solubility

The solubility of a protein depends strongly on the composition of the lysis buffer. Using the procedure described below, the solubility of a specific protein can be tested under neutral (Buffer A), high salt (Buffer B), and with detergent included (Buffer C).

1. Follow Section 3.7.1 for the best expressing condition and collect four induced samples.

2. Spin down the cells for 5 min at 12,000× g in a microcentrifuge.

3. To each cell pellet, add 100 μL of the appropriate buffer (see Section 2.10).

4. Vortex to resuspend the cells.

5. Lyse cells using the freeze-thaw method (Section 3.8.1).

6. To the supernatant, add 25 μL of 4X SDS-PAGE loading buffer.

7. To the cell pellet, add 125 μL of 1X SDS-PAGE loading buffer.

8. Heat all samples to 95°C for 5 min.

9. Vortex briefly and then centrifuge for 5 min at maximum speed.

10. Load 20 μL on an SDS-PAGE gel; avoid disturbing the pellet.

3.11 Large-scale expression of recombinant proteins

1. Transform plasmid into an *E. coli* expression strain following Section 3.2.
2. Inoculate a liquid LB culture for an overnight growth following Section 3.4.
3. The next day, use the overnight culture to inoculate 1 L of LB with the appropriate antibiotic.
4. Grow cultures at 37°C and 250 rpm shaking until the OD_{600} is ~0.6–0.8.
5. Induce expression of protein by adding IPTG to a final concentration of 0.1 mM.
6. Lower the temperature to 18°C and continue 250 rpm shaking for 12–16 h.
7. Follow Sections 3.7.1 and 3.7.2 to test for protein expression.
8. Harvest the cells by centrifugation at 6000× g.
9. Suspend cells in lysis buffer and store at −20°C.

3.12 Uniform $^{15}N/^{13}C$ labeling of recombinant proteins

This protocol is for proteins expressed under the control of the lac, tac, or T7 promoters.

3.12.1 Day 1

1. Transform 10 μL of competent BL21(DE3) cells (or derivatives) with 10 ng of plasmid DNA and plate cells on LB-agar containing the appropriate antibiotics (See Section 3.2).

3.12.2 Day 2

1. Prepare 50 mL of unlabeled defined medium for overnight culture as follows, in a 200 mL culture flask:
    - 5 mL 10X M9 medium.
    - 5 mL 10X ammonium chloride.
    - 0.75 mL 20% glucose.
    - 50 μL of each $CaCl_2$, $MgSO_4$, thiamine and biotin.
    - antibiotic at working concentration.
    - autoclaved water to 50 mL.
2. Inoculate a 5 mL culture (LB with appropriate antibiotic) with several freshly grown colonies (ca. 10–20).

3. Incubate cells in tilted tubes for a few hours at 37°C and 250 rpm in a shaking incubator, until the culture is visibly turbid.

4. Prewarm 50 mL of unlabeled defined medium to 30°C. While warming, centrifuge the LB culture (5 min, 4000× g, 30°C) and discard supernatant.

5. Resuspend cell pellet in 50 mL unlabeled defined media, for a starting OD_{600} of ~0.03–0.08. Grow the culture overnight at 30°C in a shaking incubator.

3.12.3 Day 3

1. Prepare 500 mL of ^{13}C, ^{15}N labeled defined medium as follows, in a 2 L baffled flask:

    - 50 mL 10X M9 medium.
    - 0.5 g $^{15}NH_4Cl$ dissolved in 5 mL water and sterile filtered.
    - 1.5 g ^{13}C glucose dissolved in 10 mL water and sterile filtered.
    - 500 μL of each $CaCl_2$, $MgSO_4$, thiamine and biotin.
    - antibiotic at working concentration.
    - autoclaved water to 500 mL.

2. Prewarm the 500 mL of ^{13}C, ^{15}N labeled defined medium.

3. Centrifuge the overnight 50-mL unlabeled defined medium culture (5 min, 4000× g, 30°C) and discard supernatant.

4. Resuspend the cell pellet in 500 mL of ^{13}C, ^{15}N labeled defined medium, for a starting OD_{600} of 0.03–0.08.

5. Grow culture at 37°C and 250 rpm in a shaking incubator until cells reach mid-log growth (OD_{600} ~ 0.5–0.8).

6. Once cells reach mid-log growth (OD_{600} ~ 0.5–0.8), measure the OD_{600}. Calculate the corrected volume (in mL) to take for the sample aliquot equivalent of 1 mL of cells at OD_{600} = 0.8 (See Section 3.7.1 for details).

7. Transfer aliquot to a microcentrifuge tube, and spin it down at maximum speed for at least 1 min at room temperature. Remove the supernatant. This is an uninduced sample. Store the uninduced cells at −20°C.

8. Induce protein expression by adding IPTG based on the optimal values of IPTG concentration, incubation time and incubation temperature (See Section 3.7).

9. After the induced cells have grown for the proper length of time, dilute 200 μL of the culture 10-fold with 1X PBS and measure the OD_{600}. To prepare an induced sample, take an aliquot containing the equivalent of 1 mL of cells at OD_{600} = 0.8 and immediately process it as described in Section 3.7.2.

10. Harvest the cells by centrifugation at 6000× g for 20–30 min at 4°C. Discard supernatant. Store the pellet at −20°C until ready for cell lysis.

3.13 Protein purification using IMAC

1. Resuspend cell pellet in ~35 mL of lysis buffer containing AEBSF, a protease inhibitor cocktail, and 1 mg/mL lysozyme.

2. Lyse cells (See Section 3.8).

3. Remove 75 μL of lysate and pipette into a 1.5 mL microcentrifuge tube. Centrifuge the 75 μL aliquot for 10 min at 12,000× g at room temperature.

4. Separate the supernatant into a new vial and treat with 25 μL of 4X SDS PAGE sample buffer. To the remaining pellet, add 100 μL of 1X SDS PAGE.

5. Boil separated and SDS buffer-treated samples for 10 min and store at room temperature for further SDS-PAGE analysis.

6. Centrifuge the remaining lysate (ca. 35 mL) at 16,000× g at 4°C for 20–30 min.

7. Filter the supernatant over a 0.4-micron syringe filter.

8. Pre-equilibrate the appropriate amount of Ni-NTA agarose for the amount of protein expressed in desired equilibration buffer; typically, 1–2 mL of settled agarose washed with two column volumes (CVs) of sterile, deionized water followed by two CVs of the buffer.

9. Bind the filtered lysate to the Ni-NTA agarose either batch or column loading. For batch loading, combine the filtered lysate and Ni-NTA agarose and gently rock for 30–60 min prior to pouring into the column. For column loading, pack Ni-NTA agarose into the column and pass the filtered lysate through the column. Collect the flow through eluent for SDS-PAGE analysis.

10. Wash the column with 15 CVs of cold lysis buffer. Collect wash eluent for SDS-PAGE analysis.

11. A step gradient consisting of 15 CVs each of 5, 10, and 20 mM imidazole may be used to determine best washing conditions. Collect wash eluents for SDS-PAGE analysis.

12. Wash the column with 20 CVs elution buffer, collecting 1 mL fractions.

13. Evaluate all collected samples by SDS-PAGE (see Section 3.9)[7] (**Figure 3**).

[7] Store lysate cell pellet at −20°C until SDS-PAGE has confirmed that full extraction of desired protein from the pellet is accomplished. Keep all buffers and protein samples at 4°C during purification. Batch vs. column choice will depend on binding properties of the individual protein. The SDS-PAGE of the step gradient imidazole washes will illustrate what is the highest imidazole concentration that can be used as an initial wash to clean off non-binding contaminants. If large amounts of protein remain in the cell pellet, alternative growing methods, such as IPTG concentration adjustment, or alternative purification methods including purification under denaturing conditions, should be considered. Additional purification may be necessary, such as ion exchange, or heparin binding column chromatography.

14. Pool fractions containing pure recombinant protein and dialyze into an appropriate buffer.

15. Clean Ni-NTA agarose by washing with 0.5 M NaOH for 30 min. Wash with five CVs sterile deionized water and store in 30% ethanol for long-term storage. The Ni-NTA agarose may be re-used for the same protein multiple times.

3.14 Differential scanning fluorimetry to assess protein stability

1. Prepare 1500 μL of 18 μM protein in dilution buffer.

2. Add 1.5 μL of 5000X dye.

3. Pipette up and down gently to mix.

4. Divide the protein plus dye solution among 10 vials: 140 μL per vial (some stock solution will remain).

5. Add 80 μL additive to be screened to one of nine vials.

6. Add 80 μL of dilution buffer in the tenth vial as a control.

7. Pipette up and down gently to mix.

8. Transfer 50 μL of protein plus dye plus additive solution (or control solution) to the 96-well PCR plate in triplicate.[8]

9. Cover PCR plate with a sheet of optically clear adhesive and seal each well.

10. Spin 96-well PCR plate in a centrifuge equipped with a plate holder at 800× g for 2 min at room temperature.

11. Place 96-well PCR plate into qPCR machine and run the following program:

    - select total volume per well 50 μL
    - select experiment type melting curve
    - set the following temperatures: an initial 2 min hold at 25°C, increase in increments of 0.5–1.0°C and hold each for 30 s,[9] to a final temperature of 95°C (with a 2 min hold).

12. Export data for further analysis.

13. Data can be plotted as the fluorescence vs. temperature (**Figure 4**).

14. After buffer optimization, structural data can be collected such as a ^{1}H, ^{15}N 2D NMR spectrum (see **Figure 5** for example of spectrum).

---

[8] Excess solutions are suggested to account for loss due to transfers and sticking to the sidewall of the tubes.

[9] Slower ramp rates will provide better melting temperature resolution, however not all instruments are equipped with fine temperature resolution.

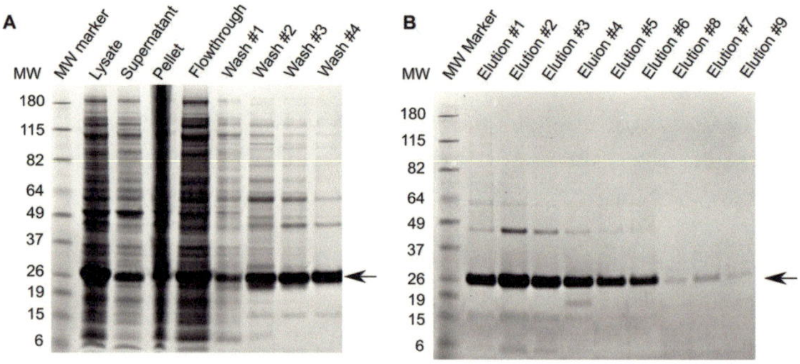

Figure 3.
SDS-PAGE gel of a typical Ni-NTA purification of an RBP (arrow indicates the recombinant RBP). (A) Samples appear in the following order: MW markers, Lysate, Supernatant, Pellet, Flowthrough, Wash #1: 50 mM Tris-Cl, 100 mM NaCl, pH 7.7; washes #2–4: 10 mM Imidazole, 50 mM Tris-Cl, 100 mM NaCl, pH 7.7. (B) MW markers, Elutions #1–9: 200 mM Imidazole, 50 mM Tris-Cl, 100 mM NaCl, pH 7.7. Some protein elutes from the column in the wash steps. All fractions are kept and can be pooled after SDS-PAGE analysis.

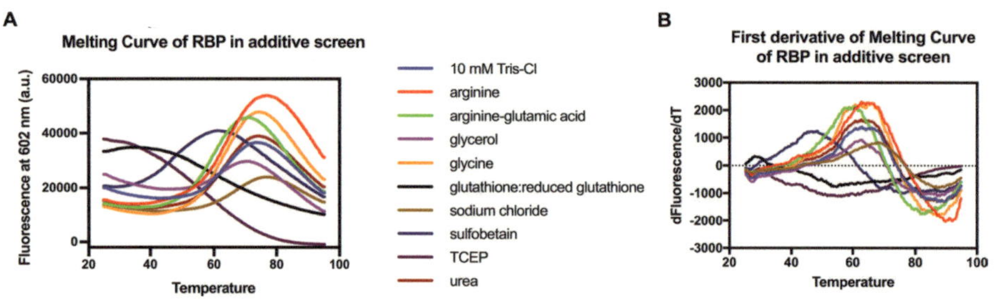

Figure 4.
DSF analysis of an RBP in buffer (10 mM Tris-Cl) with different additives. (A) A graph of the fluorescence at 602 nm at increasing temperatures for the surveyed additive screen. The inflection point preceding the peak is the melting temperature. (B) A first derivative plot with a four-point smoothing applied helps to visualize the melting temperature, where the peak is the melting temperature. The legend provides a key for both A and B.

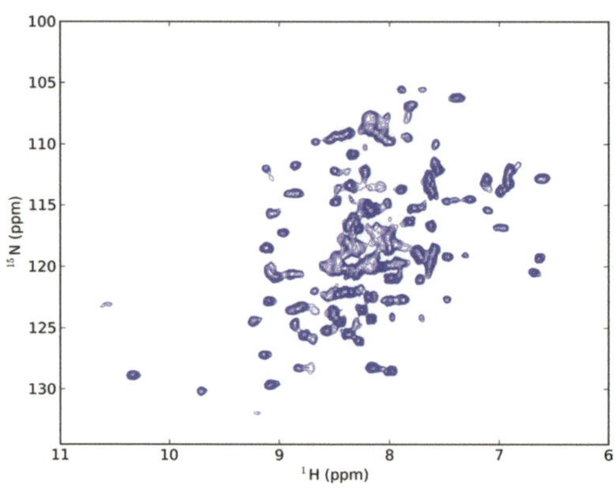

Figure 5.
^{1}H, ^{15}N 2D NMR spectrum of an RBP prepared using the methods described here.

4. Conclusion

We have described the workflow for protein expression and purification used in our shared core laboratory. These methods for growing and handling bacterial cultures work well for plasmid amplification, mini-expression screening, optimized larger-scale protein production, protein isolation and purification, and characterization of optimized experimental solution buffer conditions. Future methods can be added as needed by the users of the core and the university research community.

Acknowledgements

This publication was made possible by Institutional Development Awards (IDeA) from the National Institute of General Medical Sciences of the National Institutes of Health under Grants P20GM109095 and P20GM103408. The authors wish to acknowledge Jackson Wall for careful reading and suggestions.

Conflict of interest

Authors have no conflict of interest.

Author details

Lisa R. Warner, Olga Mass, Nancy Donnelly Lenn, Briana R. Grantham and Julia Thom Oxford*
Biomolecular Research Center, Boise State University, Boise, Idaho, USA

*Address all correspondence to: joxford@boisestate.edu

References

[1] Vincentelli R, Romier C. Expression in *Escherichia coli*: Becoming faster and more complex. Current Opinion in Structural Biology. 2013;**23**(3):326-334

[2] Rosano GL, Ceccarelli EA. Recombinant protein expression in *Escherichia coli*: Advances and challenges. Frontiers in Microbiology. 2014;**5**:172

[3] Vincentelli R, Bignon C, Gruez A, Canaan S, Sulzenbacher G, Tegoni M, et al. Medium-Scale Structural Genomics: Strategies for Protein Expression and Crystallization. Accounts of Chemical Research. 2003;**36**(3):165-172

[4] Berrow NS, Büssow K, Coutard B, Diprose J, Ekberg M, Folkers GE, et al. Recombinant protein expression and solubility screening in *Escherichia coli*: A comparative study. Acta Crystallogr Sect D Biol Crystallogr. 2006;**62**(10):1218-1226

[5] Muchmore DC, McIntosh LP, Russell CB, Anderson DE, Dahlquist FW. Expression and nitrogen-15 labeling of proteins for proton and nitrogen-15 nuclear magnetic resonance. Methods in Enzymology. 1989;**177**:44-73

[6] Fiaux J, Bertelsen EB, Horwich AL, Wüthrich K. Uniform and residue-specific 15N-labeling of proteins on a highly deuterated background. Journal of Biomolecular NMR. 2004;**29**(3):289-297

[7] Skrisovska L, Schubert M, Allain FH-T. Recent advances in segmental isotope labeling of proteins: NMR applications to large proteins and glycoproteins. Journal of Biomolecular NMR. 2010;**46**(1):51-65

[8] Tugarinov V, Kay LE. Ile, Leu, and Val methyl assignments of the 723-residue malate synthase G using a new labeling strategy and novel NMR methods. Journal of the American Chemical Society. 2003;**125**(45):13868-13878

[9] Freiburger L, Sonntag M, Hennig J, Li J, Zou P, Sattler M. Efficient segmental isotope labeling of multi-domain proteins using Sortase A. Journal of Biomolecular NMR. 2015;**63**(1):1-8

[10] Edwards AM, Arrowsmith CH, Christendat D, Dharamsi A, Friesen JD, Greenblatt JF, et al. Protein production: Feeding the crystallographers and NMR spectroscopists. Nature Structural Biology. 2000;**7**:970-972

[11] Yee A, Gutmanas A, Arrowsmith CH. Solution NMR in structural genomics. Current Opinion in Structural Biology. 2006;**16**(5):611-617

[12] Busso D, Thierry JC, Moras D. The Structural Biology and Genomics Platform in Strasbourg: An Overview. Methods in Molecular Biology. New York, NY: Humana Press; 2008. pp. 523-536

[13] Gräslund S, Nordlund P, Weigelt J, Hallberg BM, Bray J, Gileadi O, et al. Protein production and purification. Nature Methods. 2008;**5**(2):135-146

[14] Reinhard L, Mayerhofer H, Geerlof A, Mueller-Dieckmann J, Weiss MS. Optimization of protein buffer cocktails using Thermofluor. Acta Crystallographica. Section F, Structural Biology and Crystallization Communications. 2013;**69**(Pt 2):209-214

[15] Niesen FH, Berglund H, Vedadi M. The use of differential scanning fluorimetry to detect ligand interactions that promote protein stability. Nature Protocols. 2007;**2**(9):2212-2221

[16] Pantoliano MW, Petrella EC, Kwasnoski JD, Lobanov VS, Myslik J,

Graf E, et al. High-density miniaturized thermal shift assays as a general strategy for drug discovery. Journal of Biomolecular Screening. 2001;**6**(6):429-440

[17] Senisterra GA, Markin E, Yamazaki K, Hui R, Vedadi M, Awrey DE. Screening for ligands using a generic and high-throughput light-scattering-based assay. Journal of Biomolecular Screening. 2006;**11**(8):940-948

[18] Nettleship JE, Brown J, Groves MR, Geerlof A. Methods for Protein Characterization by Mass Spectrometry, Thermal Shift (ThermoFluor) Assay, and Multiangle or Static Light Scattering. New York, NY: Humana Press; 2008. pp. 299-318

[19] Ericsson UB, Hallberg BM, DeTitta GT, Dekker N, Nordlund P. Thermofluor-based high-throughput stability optimization of proteins for structural studies. Analytical Biochemistry. 2006;**357**(2):289-298

[20] Vedadi M, Niesen FH, Allali-Hassani A, Fedorov OY, Finerty PJ, Wasney GA, et al. Chemical screening methods to identify ligands that promote protein stability, protein crystallization, and structure determination. Proceedings of the National Academy of Sciences of the United States of America. 2006;**103**(43):15835-15840

[21] Fallahi A, Kroll B, Warner LR, Oxford RJ, Irwin KM, Mercer LM, et al. Structural model of the amino propeptide of collagen XI alpha1 chain with similarity to the LNS domains. Protein Science. 2005;**14**(6):1526-1537

[22] Warner LR, Fallahi A, Kroll B, Irwin KM, Yingst S, Mercer LM, et al. Modeling and characterization of the amino propeptide of collagen α1(XI), a regulatory domain in collagen fibrillar architecture. Materials Research Society Symposium Proceedings, Structure and Mechanical Behavior of Biological Materials. 2005;**874**:41-46

[23] Oxford JT, DeScala J, Morris N, Gregory K, Medeck R, Irwin K, et al. Interaction between amino propeptides of type XI procollagen alpha1 chains. The Journal of Biological Chemistry. 2004;**279**(12):10939-10945

[24] Medeck RJ, Sosa S, Morris N, Oxford JT. BMP-1-mediated proteolytic processing of alternatively spliced isoforms of collagen type XI. The Biochemical Journal. 2003;**376**(pt 2):361-368

[25] Warner LR, Brown RJ, Yingst SM, Oxford JT. Isoform-specific heparan sulfate binding within the amino-terminal noncollagenous domain of collagen α1(XI). The Journal of Biological Chemistry. 2006;**281**(51):39507-39516

[26] Ryan RE, Martin B, Mellor L, Jacob RB, Tawara K, McDougal OM, et al. Oncostatin M binds to extracellular matrix in a bioactive conformation: Implications for inflammation and metastasis. Cytokine. 2015;**72**(1):71-85

[27] Kahler RA, Yingst SMC, Hoeppner LH, Jensen ED, Krawczak D, Oxford JT, et al. Collagen 11a1 is indirectly activated by lymphocyte enhancer-binding factor 1 (Lef1) and negatively regulates osteoblast maturation. Matrix Biology. 2008;**27**(4):330-338

[28] Oxford JT, DeScala J, Morris N, Gregory K, Medeck R, Irwin K, et al. Interaction between amino propeptides of type XI procollagen α1 chains. The Journal of Biological Chemistry. 2004;**279**(12):10939-10945

[29] Gregory KE, Oxford JT, Chen Y, Gambee JE, Gygi SP, Aebersold R, et al. Structural organization of distinct domains within the non-collagenous N-terminal region of collagen type XI. The Journal of Biological Chemistry. 2000;**275**(15):11498-11506

Chapter 3

Carbapenemases

Branka Bedenić and Sanda Sardelić

Abstract

Carbapenems are usually regarded as the last treatment option for serious infections caused by Gram-negative and Gram-positive microorganisms. Although they are stable to hydrolysis by most β-lactamases, their usage as the last resort antbiotics was seriously compromised by the appearance of carbapenem-hydrolyzing enzymes called carbapenemases. These enzymes are produced mostly by *Enterobacteriaceae* and Gram-negative nonfermentative bacteria such as *Pseudomonas aeruginosa* and *Acinetobacter baumannii*. True carbapenemases belonging to Ambler molecular classes A, B, and D are often encoded by genes embedded in mobile genetic elements like plasmids, integrons, and transposons, which often harbor multiple resistance determinants limiting further the treatment options. At present, large number of nosocomial and community-acquired infections caused by worldwide spread of carbapenem-resistant Gram-negative bacteria have become a major public health problem. Although polymyxins remain active, in vitro reports of benefits of combination schemes favor this strategy against carbapenemase-producing Gram-negative bacilli.

Keywords: carbapenems, carbapenemases, *Enterobacteriaceae*, *Acinetobacter baumannii*, *Pseudomonas aeruginosa*

1. Introduction

Carbapenems are bactericidal antibiotics often used as a last treatment option for severe infections caused by isolates producing extended-spectrum and/or AmpC β-lactamases. However, their stability to β-lactamases of Gram-negative bacteria is not absolute. From the first report of carbapenemases in the 1980s, the last decade is marked with the rapid spread of carbapenemases-producing organisms, becoming a public health concern worldwide.

2. Carbapenem antibiotics

Carbapenems are β-lactam antibiotics with broad antimicrobial spectrum, which encompasses Gram-positive and Gram-negative aerobic organisms and anaerobic bacteria. They often remain as the last treatment option for infections caused by multiresistant Gram-negative isolates. Their in vitro efficacy is due to good penetration through the outer membrane of Gram-negative bacteria, high affinity for penicillin-binding protein (PBP) molecules and excellent stability against β-lactamases [1]. The carbapenems bind to PBP-1 and PBP-2 and induce spheroplast formation and cell lysis without filament formation. They comprise imipenem, meropenem, ertapenem, and doripenem. In general, imipenem, and

IntechOpen

doripenem are potent also against Gram-positive bacteria, while meropenem, ertapenem, and doripenem are slightly more effective against Gram-negative organisms. Ertapenem is not active against *P. aeruginosa* or *A. baumannii*, while doripenem has lower minimum inhibitory concentrations (MICs) than imipenem and meropenem versus *P. aeruginosa* and *A. baumannii* [1] Meropenem has an advantage over imipenem because it is not susceptible to renal dihydropeptidase and can be administered without cilastatin which is neurotoxic and contraindicated in neurological patients. Carbapenems are stable against wide range of serine-based β-lactamases including extended-spectrum β-lactamases (ESBLs), plasmid-mediated AmpC β-lactamases or against strains with inducible and derepressed expression of chromosomal AmpC β-lactamases. Although the stability against these common enzymes is not complete, the rate of hydrolysis is slow and only becomes significant when coupled with reduced outer membrane permeablity. Carbapenems are strong inducers of chromosomal β-lactamases, but β-lactamase induction does not affect their activity [2]. However, induction of chromosomal AmpC beta-lactamase can cause antagonism with penicillins and cephalosporins, and thus carbapenems should not be administed in combination with other β-lactams.

3. Classification, historical background, and properties of carbapenemases

Some β-lactamases, named carbapenemases, can efficiently hydrolyze carbapenems [3]. These include chromosomal enyzmes, which are ubiquitious in some unusual pathogens and mostly environmental isolates (e.g., *Stenotrophomonas maltophilia*, *Aeromonas* spp., *Flavobacterium* spp., *Bacteroides fragilis*) and acquired carbapenemases of classes A, B, and D [3]. Although several class C carbapenemases have been described (ACT-1, CMY-2, CMY-10, and ADC-68), their producers usually exhibit reduced susceptibility to carbapenems as a consequence of low enzyme's catalytic efficiency and a permeability defect. Therefore, they are not considered as true carbapenemases [4].

Based on the structure of their active site, carbapenemases are devided into two groups: class A-penicillinase and class D-oxacillinases; they contain serine at their active site and are inhibited by β-lactam inhibitors including clavulanic acid and tazobactam. The second group is metallo-β-lactamases belonging to class B which contain zinc atom at the active site and is inhibited by metal chelators such as ethylenediaminetetraacetic acid (EDTA) [5].

3.1 Class A carbapenemases

Serin carbapenemases of molecular (Ambler) group A are rare, detected mainly in *Enterobacteriaceae*. They belong to the functional subgroup 2f according to the classification scheme proposed by Bush et al. [6, 7]. The genes encoding them are located on either chromosome or plasmid. This group comprises SME-1 (*Serratia marcescens)* [8], SME-2 (*S. marcescens)* [9], SME-3 (*S. marcescens)* [10], IMI-1 (*Enterobacter cloacae)* [11], NMC-A (*E. cloacae)* [12], KPC-1, KPC-2, and KPC-3 (*Klebsiella pneumoniae*) [13–15] and GES-2 (*P. aeruginosa)* [16]. They confer resistance to aminopenicillins, ureidopenicillins, first- and second-generation cephalosporins, aztreonam, and imipenem. Imipenem is hydrolyzed very slowly without causing a clinically relevant resistance [17]. SME β-lactamase was reported for the first time in *S. marcescens* in the UK in 1982 [8]. Infections caused by SME-positive *S. marcescens* are usually sporadic without potential of causing outbreaks. IMI and NMC β-lactamases have been sporadically detected in isolates of *E. cloacae* in the USA, France, and Argentina [9, 10]. These β-lactamase are encoded chromosomally and have never been described on transferable genetic elements explaining their sporadic reports wordwide [17]. KPC-1 is a novel class A enzyme isolated in 1996 from *K. pneumoniae* in the USA [13]. The discovery of KPC-1 was quickly followed by two new allelic variants: KPC-2 and KPC-3 [14]. Subsequent revision of the bla_{KPC-1} sequence demonstrated that KPC-1 and KPC-2 are identical enzymes [13–15]. KPC-3 differs from KPC-2 (the former KPC-1) by one amino acid change (H272Y) [14]. Although, so far, more than 20 different KPC variants have been described, KPC-2 and KPC-3 remain the most commonly identified. Unlike other class A carbapenemases, they hydrolyze expanded-spectrum cephalosporins and all carbapenems. KPC β-lactamases are weakly inhibited by clavulanic acid and tazobactam and are often coharbored with OXA-1 conferring resistance to β-lactam inhibitor combinations. They are not inducible and are encoded on transferable plasmids. A 10-kb Tn3-type transposon Tn*4401* with two insertion sequences IS*Kpn6* and IS*Kpn7* is repeatedly reported as the main genetic structure enhancing the spread of bla_{KPC}-type genes [5]. Recently, a chromosal integration of the bla_{KPC} gene was shown in *K. pneumoniae*, thus enabling easier maintenance of the gene in bacterial population [18]. Today, KPC enzymes are highly prevalent, not only found in *K. pneumoniae* but also in an increasing number of *Enterobacteriaceae, P. aeruginosa*, and *A. baumannii*. They are not only found all over the USA but also in Central and South America, while in European countries, KPC-producing isolates are mostly prevalent in Mediterranean countries, especially in Italy and Greece, countries reported as having endemic situation [19, 20].

They are also still present in Middle East where in 2007 a nationwide strategy was implemented to contain the outbreak faced by Israeli hospitals [21].

KPC-producing isolates are also reported in Asia, especially China where a hypervirulent *Klebsiella pneumoniae* strain K1-producing KPC is concerning [22, 23].

The main transporter in the worldwide dissemination of KPC-producing enzymes is succesfull clone *Klebsiella pneumoniae* ST258 [24]. In contrast to the majority of other class A carbapenemases, KPCs are typically plasmid mediated and usually carried on IncF plasmid. Plasmid location of bla_{KPC} genes is responsible for the rapid dissemination of those important resistance determinants [23, 24].

GES-2 β-lactamase was derived by point mutation from GES-1 extended-spectrum β-lactamase (with amino acid change G170 N) which was described in *K. pneumoniae* isolate from French Guinea [16, 25]. GES-2 β-lactamase was first reported from *Pseudomonas aeruginosa* from South Africa, was plasmid mediated, and conferred resistance or reduced susceptibility to carbapenems [26]. Because the enzymes have broad hydrolysis spectrum that includes penicillins and expanded-spectrum cephalosporins, they were initially classified as ESBLs. Their hydrolysis spectrum was expanded in 2001 to include imipenem. GES β-lactamases are related to integron-borne-cephalosporinase—IBC) ESBLs, but confusing IBC names have been converted to GES [2, 3, 17]. Genes encoding GES family of β-lactamases are located on integrons embedded in plasmids with a possibility of conjugal transfer. However, they appear rarely in contrast to KPC β-lactamases and are not frequently associated with nosocomial outbreaks [3, 17, 26]. At present, the number of GES variants is increasing (with more then 20 described), but only several of them (mainly GES-2, -6, -14, -20) display carbapenemase activity (others are mostly ESBL). GES-type carbapenemases were identified in *Pseudomonas aeruginosa*, in *Acinetobacter baumannii*, and *Enterobacteriaceae* as well (mostly *Klebsiella pneumoniae*, *Serratia marcescens*, and *Enterobacter* spp. isolates) and have worldwide distribution [17, 23]. In spite of the fact that they are plasmid mediated, they do not have a capacity for successful dissemination like KPC [23].

3.2 Class B carbapenemases

Metallo-β-lactamases (MBLs) are clinically the most relevant carbapenemases [23, 27]. This group of enzymes catalyzes the β-lactam hydrolysis by active site containing one or more bounded zinc ions that promote formation of nucleophilic hydoxide further attacking the β-lactam ring. The first descriptions of metallo-β-lactamases included studies on environmental and opportunistic bacteria such as *Stenotrophomonas maltophilia*, *Bacillus cereus*, *Aeromonas* spp., *Legionella gormanii*, *Pseudomonas stutzeri, Shewanella* spp., *Myroides odoratus*, and so on, which carry MBLs on their chromosome but could present potential reservoir of mobile carbapenemase determinants [27, 28].

The metallo-β-lactamases or class B carbapenemases are further divided into subclasses (B1, B2 and B3), but the largest number of clinically relevant MBLs belong to B1 subclass, including the most frequently described Verona integron-encoded MBL (VIM), imipenemase (IMP), and New Delhi MBL (NDM). Those MBLs are usually located within different integron structures, and these integrons are associated with mobile plasmids or transposons facilitating the transfer of resistance genes between bacteria [3–5]. Acquired MBLs are reported from all over the world but mostly from Europe and Far East [27]. Transferable imipenem resistance was first detected in 1990 in Japan in *P. aeruginosa* isolate [29]. The β-lactamase-conferring imipenem resistance was designated IMP-1 and was located on a conjugative plasmid [30]. The same enzyme was later found on an integron in *S. marcescens* and other *Enterobacteriaceae* also in Japan. The first member of IMP family reported in Europe was IMP-1 found in *A. baumannii* in Italy [31]. The gene encoding this enzyme was located on class 1 integron [32]. Since that time, the IMP family was found throughout the world, but most reports originate from Far East [4, 22, 27, 33]. IMP-2 β-lactamase was found in an *Acinetobacter baumanni* isolate from Italy in 1997 and shares 85% of amino acid identity with IMP-1 with similar kinetic parameters, while IMP-3 was reported in *Shigella flexneri* from Japan. That was the first appearance of MBL in typical community-acquired pathogen [34]. IMP-3 enzyme was previously known as MET-1, but the substrate specificity, that is, preferential hydrolysis of cephalosporins (and not penicilins and carbapenems), was a direct consequence of a single-base mutation (although the genes differ in two bases and two amino acid changes, that is, S262G and E126G). The gene and the enzyme were then renamed to bla_{IMP-3} or IMP-3 β-lactamase. Since IMP-1 shows broader substrate profile, it seemed that the gene blaIMP-3 was an ancestor of bla_{IMP}-1 [27]. IMP-4 MBL was first reported from *A. baumannii* associated with nosocomial outbreak in Hong Kong in 2001 [35]. IMP-5 was identified in urinary isolate of *A. baumannii* from Portugal in 2002, highly resistant to penicillins, broad-spectrum cephalosporins, including ceftazidime, ceftriaxone, cefepime, cefpirome, and to aztreonam, but remained susceptible to ampicillin/sulbactam, aminoglycosides, and quinolones [36]. IMP-5 showed greater homology with IMP-1, IMP-3, and IMP-4 than with IMP-2. $Bla_{\mathrm{IMP-5}}$ was the only gene cassette inserted into a class 1 integron, named In76. This was the first IMP-β-lactamase in Portugal and the second in Europe [36]. There are more than 50 allelic IMP variants reported so far [37]. They are usually devided into IMP-1 and IMP-2 cluster, suggesting their different phylogenetic origin. IMP-12, a very different variant outside the IMP-1 cluster (it has 85% similarity to IMP-1), has a unique structural feature and shows reduced catalytic effieciency toward penicillins [38]. IMP-51 was recently identified in a *P. aeruginosa* clinical isolate in Vietnam and showed increased doripenem- and meropenem-hydrolyzing activities

[39]. IMP-type enzymes have been reported all over the world in *Enterobacteriaceae*, *P. aeruginosa*, and *A. baumannii* but still represent the dominant carbapenemases in Japan and in Southeast Asia [33, 38, 40]. Analysis of the genetic platform of *bla*$_{IMP}$ genes has shown that most of them are embedded as gene cassettes in class I integron, harboring also other resistance genes, such as aac (mediating resistance to aminoglycosides), *bla*$_{OXA}$ (different serine oxacillinases), resistance to antiseptics, or chloramphenicol [38, 40, 41].

Another prevalent family of integron-associated metallo-β-lactamases is composed of VIM enzymes. VIM means Verona integron-encoded metallo-β-lactamase because the first enzyme in this family was reported in *Pseudomonas aeruginosa* isolate from Verona in 1997 [42]. *Bla*$_{VIM-1}$ gene was located within class 1 integron repeatedly showing the important role in the dissemination of IMP and VIM genes not only in nonfermentative bacteria but also in *Enterobacteriaceae* [4]. VIM-2 was described in *Pseudomonas aeruginosa* from France that was isolated in 1996 and shared 90% similarity with VIM-1 [43]. VIM-3 was reported in *P. aeruginosa* isolates associated with nosocomial outbreak from Taiwan [44]. The enzyme was chromosomally encoded and differed in two amino acids from VIM-1, namely C178A and A443G. VIM-4 β-lactamases were identified in *K. pneumoniae* and *Enterobacter cloacae* isolates from Italy in a patient previously treated with carbapenems and has one amino acid change in comparison to VIM-1 (S228A), being more active against imipenem than VIM-1 [45]. There are more than 40 allelic variants of VIM enzymes reported so far, and they mainly belong to three sublineages—VIM-1-like, VIM-2-like, and VIM-7-like enzymes, representing phylogenetically the most divergent variants [38, 46]. VIM-2-like enzymes have been associated mostly with *P. aeruginosa*, whereas VIM-1-like enzymes, in particular, VIM-4, have been frequently reported in strains of the *Enterobacteriaceae* [38, 40, 46]. It is interesting to mention that if VIM-1 or VIM-4 enzymes are found in *E. cloacae* and *K. pneumoniae*, mostly of ST147 and 11, they are usually carried on incompatibility group N or A/C plasmids, whereas in *E. coli*, those genes are often associated to IncFI/II plasmids [38, 41]. VIM-7, a more divergent enzyme, with lower efficiency for cephalosporins, was for the first time reported in the USA in 2001, and it was the first MBL in North America [47, 48]. North America was spared from MBLs until early 2000s. VIM enzymes are today widely distributed in all continents and at present the most common MBL found all over the world. In Mediterranean countries, such as Italy and Greece, VIM-positive *P. aeruginosa* strains of clonal complex 235 and 111 have been involved in nosocomial outbreaks [38]. The activity of aztreonam, as the only β-lactam not hydrolyzed by MBLs, is strongly compromised by the copresence of ESBLs, plasmid-mediated AmpC, or other serine β-lactamase together with MBLs [49, 50]. *Bla*$_{VIM}$ genes are typically embedded in class 1 integrons, which can be incorporated in either plasmids or chromosome. Plasmids carrying *bla*$_{VIM}$ genes in *Enterobacterriaceae* belong most frequently to IncA/C or IncN group [23, 38, 41].

NDM is a new family of metallo-β-lactamases unrelated to other MBLs. The first NDM-1 positive isolate from Europe was reported in Sweden from a patient who contracted a *K. pneumoniae* infection in India in late 2007 [51]. At the beginning, NDM enzymes spread in the Indian subcontinent, and they are still endemic in India, Pakistan, and Bangladesh. They were found not only in clinical isolates but also in public tap water and sewage water [52].

The spread soon involved the UK, historically related to India, but until now, it was described as present in all continents. However, it should be noted that Balkan region, Arabian Peninsula, and North African countries are considered as possible additional reservoirs of NDM-producing strains [38, 53, 54].

So far, the NDM enzymes have been found mostly in *Enterobacteriaceae* (with the importance of epidemic *Klebsiella pneumoniae* ST147, 11 and 14 and *E. coli*

ST131 and 405), *P. aeruginosa*, and *A. baumannii*. *Enterobacteriaceae*-producing NDM β-lactamases pose a serious public health concern because huge conjugative plasmids carrying the bla_{NDM} genes can have up to 14 other antibiotic resistance determinants and can trasfer this resistance to other bacteria, resulting in multidrug resistance or pan-drug resistance phenotypes [55].

The genetic platform of bla_{NDM} gene mostly involves transposon Tn*125* with two flanking IS*Aba125* elements, often truncated in *Enterobacteriaceae* and part of broad-host-range plasmid including IncA/C (predominant plasmid type), F, R, H, N, L/M, and X types [38, 41, 53, 54]. NDM producers can acquire additional β-lactamases, even carbapenemases, and are recognized as the source of community-acquired infections [54, 55]. The highest distribution of NDM variants (more than 15 so far) is detected in *K. pneumoniae* and *E. coli* species from Asian continent with China and India still serving as the major reservoir of NDM producers [38, 54].

A global surveillance study conducted through 2012–2014 enrolled 202 medical centers from 40 countries all over the world and identified 471 MBL-producing isolates of *Pseudomonas aeruginosa* and *Enterobacteriaceae* with 32 different MBL variants. Among them, seven novel allelic MBL variants were identified (VIM-42, -43, -44 and =45, IMP-48 and -49 and NDM-16), differing by one amino acid from previously reported enzymes proving still dynamic genetic background of these worldwide disseminated enzymes [56].

Other MBL-encoding genes are mostly limited to not only a single report or to particular geographic regions, including genes encoding SPM-1, GIM-1 and SIM-1, but also more recently described AIM-1, KHM-1, DIM-1, BIC-1, and TMB-1 [38, 57–60]. GIM–MBL, unlike other class B carbapenemases, has two zinc ions in the active site.

SPM β-lactamases, originally described in *P. aeruginosa* strain from Brazil, are mostly limited to South America and have been associated with outbreaks at many hospitals. There were two imported cases in Europe until now, one in Switzerland, and one in the UK [57, 61, 62]. Unlike the other MBL-encoding genes, bla_{SPM} gene was not part of gene cassette or integron, but it was located on plasmid and flanked with two expanded form is insertion segment common region (IS*CR*) elements probably providing its mobilization capacity [57]. The gene was also described in *A. baumannii* strain from Tehran [63].

3.3 Class D carbapenemases (carbapenem hydrolyzing class D β-lactamases; chdls)

The main characteristic of these enzymes is their hydrolytic activity toward oxacillin. Ambler group D carbapenemases include oxacillinases typical for *Acinetobacter baumannii*. [64]. OXA enzymes (the name came since they hydrolyze isoxazolylpenicillins much faster than penicillins) were recently divided into 12 subgroups: OXA-23-like, OXA-24/40-like OXA-48, OXA-51-like, OXA-58-like, OXA-143-like, OXA-253, OXA-211, OXA-213, OXA-214, OXA-229, and OXA-235 [65]. The amino acid sequence identities between members of the same subgroup are more than 90%, whereas the identities between enzymes that belong to different subgroups are less than 70% [65]. Enzymes belonging to OXA-51 group are naturally occurring β-lactamases of *A. baumannii* (often used as marker of identification) and are normally expressed at low levels but can be overexpressed as a consequence of the IS*Aba1* location upstream of the genes [64]. The clinically significant resistance is usually expressed in combination with other mechanisms like porin loss or upregulated efflux pumps. OXA-23 subgroup has been reported worldwide but most reports are coming from *Acinetobacter baumannii* isolates from Latin America. OXA-23-encoding gene is usually located on transposon Tn2006, almost always bracketed on both sides by IS*Aba1* insertion sequence [5, 65–67]. It could be located on transferable plasmids, but it was also described integrated in the *A. baumannii* chromosome. OXA-24/40 subgroup was reported less frequently from *A. baumannii*, mostly from Iberian Peninsula and the USA [68]. It confers high level of carbapenem resistance and is usually chromosomally encoded. OXA-58 and OXA-143 are usually plasmid mediated, first reported in France and Germany, respectively [69, 70]. Similar to OXA-23, $bla_{\text{OXA-58}}$ is located within transposon usually bracketed by two copies of IS*Aba3*. OXA-72 is the most prevalent allelic variant found in Europe, Far East, and South America [65, 71–73]. OXA-72 and OXA-23 variants have not only been identified in long-term care facilities but also in environmental samples such as municipal wastewater stressing the fact that strains resistant to antimicrobials are constantly circulating between the hospitals, community, and the nature [74–76].

Newer oxacillinase subgroups were mostly described as naturally occurring enzymes with low carbapenemase activity in *Acinetobacter* "non-*baumannii*" isolates like OXA-134a and OXA-143 in *A. lwoffi*, OXA-211 in *A. johansonii*, OXA-213 in *A. calcoaceticus*, OXA-214 in *A. haemolyticus*, OXA-229 in *Acinetobacter bereziniae*, and OXA-235 in *Acinetobacter schindleri* [65, 66, 70]. However, it is suggested that those enzymes present a reservoir of enzyme variants that could evolve to a more significant level and could be easily transferred to other bacterial species.

OXA-48 β-lactamase was originally described in a *Klebsiella pneumoniae* isolate from Istanbul, Turkey, in 2001 [77]. It is now widespread not only in *Klebsiella pneumoniae* but also in other *Enterobacteriacea*. Turkey was reported as having the highest epidemiologic level called endemic in 2015. The most important reservoirs of OXA-48 carbapenemases are linked with Turkey, India, Middle East, and North African countries [78]. $bla_{OXA\text{-}48}$ gene is usually located in pOXA-48a plasmid, which belongs to the group of IncL/M plasmids. This plasmid has high conjugation rate, and it is self-conjugative, explaining partly the global dissemination of OXA-48 enzymes [53]. The spread of $bla_{OXA\text{-}48}$ gene is further mediated by IS*1999* insertion sequence located upstream of the Tn*1999* transposon which is usually carried on L/M plasmid. In the majority of European countries, at present, it is the most prevalent type of carbapenemase among *Enterobacteriaceae* [65, 78].

OXA-48 hydrolyzes carbapenems at a low level, shows very weak activity against expanded-spectrum cephalosporins, and does not significantly hydrolyze ceftazidime and cefepime but in combination with impermeability can lead to high-level resistance to carbapenems [79]. OXA-48 significantly hydrolyze only penicillins and narrow-spectrum cephalosporins and like other class D enyzmes is not inhibited by clavulanic acid, sulbactam, or tazobactam leading to elevated MICs of amoxycillin/clavulanate and piperacillin/tazobactam, as shown in **Table 1**. Resistance to expanded-spectrum cephalosporins occurs only in strains positive for addtional ESBL or AmpC. Some OXA-48 variants, including OXA-181 (differing by four amino acids from OXA-48) and OXA-162 (with one amino acid substitution T213A) were also found in *Enterobacteriaceae* [78]. OXA-162 variant is also able to hydrolyze ceftazidime, whereas OXA-48 is not. Interestingly, OXA-181 is often associated with other carbapenemase genes, and many reports from Indian subcontinent describe isolates coproducing OXA-181, NDM-1, and VIM enzymes [80].

In a recent report, environmental species, *Shewanella xiamenensis* was identified as the progenitor of the $bla_{\text{OXA-181}}$ gene [81]. It is therefore hypothesized that a mobilization of a chromosomal gene to plasmid happened, and then the gene disseminated further in clinically relevant species. This work emphasizes the constant interplay between microorganisms in the environment, community, and clinics. Gram-negative bacteria are ubiquitous in nature and are widely distributed in soil and water where they can survive for a long time posing a possible reservoir of

Type of carbapenemase	Substrate profile						Inhibiton by	Country of origin	Host species	Reference
	Penicillins	Narrow-spectrum cephalosporins	Expanded-spectrum cephalosporins	Aztreonam	Imipenem	Meropenem				
Class A										
IMI	+	+	—	—	+	—	Clavulanic acid	France	*E. cloacae*	[11]
NMC	+	+	—	—	+	—	Clavulanic acid	France	*E. cloacae*	[12]
SME-1	+	+	—	—	+	—	Clavulanic acid	United States	*S. marcescens*	[8]
SME-2	+	+	—	—	+	—	Clavulanic acid	United States	*S. marcescens*	[9]
SME-3	+	+	—	—	+	—	Clavulanic acid	United States	*S. marcescens*	[10]
KPC-1	+	+	+	+	+	+	PBA	United States	*K. pneumoniae*	[13]
KPC-2	+	+	+	+	+	+	PBA	United States	*K. pneumoniae*	[13, 15]
KPC-3	+	+	+	+	+	+	PBA	United States	*K. pneumoniae*	[14]
GES-2	+	+	—	—	+	—	Clavulanic acid	French Guinea	*P. aeruginosa*	[16]
Class B										
IMP-1	+	+	+	—	+	+	EDTA	Japan	*P. aeruginosa*	[29]

Type of carbapenemase	Substrate profile						Inhibiton by	Country of origin	Host species	Reference
	Penicillins	Narrow-spectrum cephalosporins	Expanded-spectrum cephalosporins	Aztreonam	Imipenem	Meropenem				
IMP-2	+	+	+	—	+	+	EDTA	Italy	***A. baumannii***	[32]
IMP-4	+	+	+	—	+	+	EDTA	Japan	***A. baumannii***	[35]
IMP-5	+	+	+	—	+	+	EDTA	Portugal	***A. baumannii***	[36]
VIM-1	+	+	+	—	+	+	EDTA	Italy	*P. aeruginosa*	[42]
VIM-2	+	+	+	—	+	+	EDTA	France	*P. aeruginosa*	[43]
VIM-3	+	+	+	—	+	+	EDTA	Taiwan	*P. aeruginosa*	[44]
VIM-4	+	+	+	—	+	+	EDTA	France	*E. cloacae, K. pneumoniae*	[45]
VIM-7	+	+	—	—	+	+	EDTA	USA	*P. aeruginosa*	[47]
NDM-1	+	+	+	—	+	+	EDTA	Sweden	*K. pneumoniae*	[51]
GIM-1	+	+	+	—	+	+	EDTA	Germany	*P. aeruginosa*	[58]
SPM-1	+	+	+	—	+	+	EDTA	Brasil	*P. aeruginosa*	[57]

Type of carbapenemase	Substrate profile						Inhibiton by	Country of origin	Host species	Reference
	Penicillins	Narrow-spectrum cephalosporins	Expanded-spectrum cephalosporins	Aztreonam	Imipenem	Meropenem				
SIM-1	+	+	+	—	+	+	EDTA	South Korea	*P. aeruginosa*	[59]
DIM-1	+	+	+	—	+	+	EDTA	The Netherlands	*P. aeruginosa*	[60]
Class D										
OXA-48-like	+	+	—	—	+	—		Turkey	*K. pneumoniae*	[77]
OXA-23 (ARI-1)	+	+	+	+	+	+		UK	***A. baumannii***	[67]
OXA-24/40-like	+	+	+	+	+	+	NaCl	Spain	***A. baumannii***	[68]
OXA-58-like	+	+	+	+	+	+		France	***A. baumannii***	[69]
OXA-143-like	+	+	+	+	+	+		Germany	***A. baumannii***	[70]
OXA-253	+	+	+	+	+	+		Germany	***A. baumannii***	[65]

The most prevalent allelic variants are shown. Abbreviations: PBA-phenylboronic acid; EDTA-ethylenediaminetetraacetic acid.

Table 1.
Classification, substrate profile, susceptibility to inhibitors, host species, and county of origin of carbapenemases.

resistance determinants. Main characteristics of the most prevalent allelic variants of the Ambler classes A, B, and D carbapenemases are shown in **Table 1**.

4. Clinical significance of carbapenemases

Carbapenemase-producing Gram-negative bacteria can cause a wide spectrum of infections including bacteremia, nosocomial pneumonia, wound infections, endocarditis, and urinary tract infections. Risk factors for carbapenem-resistant isolates usually include procedures involving manipulations of the gastrointestinal tract, trauma, orthopedic procedure, or life-threatening conditions like septicemia and septic shock. Those infections are often associated with high hospitalization cost, treatment failures, longer hospital stay, and high mortality rates [82–84].

5. Laboratory detection

Appropriate and timely laboratory detection of carbapenemase-producing microorganisms is of crucial importance to implement adequate antimicrobial therapy. Screening for carbapenemase production is usually based on reduced inhibition zones around carbapenem disks in routine disk-diffusion testing or elevated MICs for carbapenems. However, the level of resistance depends on the type of carbapenemase, allelic variant, gene expression, and/or the presence of additional resistance mechanisms such as porin loss or upregulation of efflux pumps.

Recommendations for carbapenem susceptibility breakpoints have been established by both Clinical Laboratory Standards Institute (CLSI) and The European Committee on Antimicrobial Susceptibility Testing (EUCAST), but no global consensus exists with EUCAST breakpoints being generally lower [85, 86]. However, some strains with low-level resistance are still being missed. EUCAST has also established epidemiologic breakpoints to distiguish between wildtype isolates and those with carbapenem-resistant determinants [86].

Ideally, methods for identifying carbapenemase should have a short turnaround time to ensure timely implementation of control measures. This could be challenged by difficulties in detecting carbapenemase producers, since MICs to carbapenems could be elevated but within susceptible range or even low, as described in *Enterobacteriacea*e and *A. baumannii*. Particularly in OXA-48-producing organisms, the MICs of carbapenemas can be very low. The screening usually relies on the reduction of the inhibition zone around ertapenem disk [87, 88]. However, relevant methodology with specific laboratory test has not yet been standardized. Modified Hodge test is the only test recommended by CLSI for the phenotypic detection of carbapenemase producers but often lacking sensitivity and specificity. There are also several inhibitor-based tests using different inhibitors (EDTA and phenantroline as inhibitors of MBLs, phenylboronic acid as inhibitor of KPC) in combination with carbapenem (e.g., meropenem) or cephalosporin (e.g., ceftazidime) in different formats—disk-diffusion or broth dilution or E-test [88]. There is no specific inhibitor that could be used in detection of class D carbapenemases, but there are reports on using temocillin disk (or combined with avibactam) for this purpose [88]. Carba NP (derived from the names Nordmann-Poirel) test is a simple biochemical test based on hydrolysis of imipenem detectable by a change of color of the indicator due to decrease of pH. It is applicable in most microbiological laboratories, although the reference standard in detection of carbapenemase production is spectrophotometric measurement of carbapenem hydrolysis in the presence or absence of inhibitor, but it is still reserved for reference laboratories [89]. Recently, a new method for the detection of carbapenemases was described [90].

The test was called carbapenem inactivation method (CIM), and it was based on degradation of meropenem by carbapenemase. The water suspension of carbapenem-resistant microorganism is incubated with a 10-ug meropenem disk for 2 h. After incubation, the disk is removed from the suspension and placed onto a Mueller-Hinton agar plate previously inoculated with the carbapenem-sensitive microorganism (usually *Escherichia coli*). If the carbapenem-resistant microorganism produces carbapenemase, consequently, meropenem will be hydrolyzed and the indicator (carbapenem susceptible) microorganism will grow close to the disk [90]. CIM method has high sensitivity, comparable to that of Carba NP test, as reported recently [91]. Modified CIM test includes preparation of the bacterial suspension in tryptic soy broth and extending the time of incubation to 4 hours improving the detection of some carbapenemases in *Enterobacteriaceae* [92]. Recently, the use of mass spectrometry (matrix-assisted laser desorption/ionization-time of flight—MALDI-TOFF) based on analysis of degradation of carbapenem-molecule-enabled rapid detection of KPC carbapenemase (in 45 min) or MBL (150 min) [93]. Finally, simplex or multiplex PCR, real-time PCR, or hybridization tests could significantly improve detection of carbapenemase genes in clinical laboratory bypassing the sensitivity and specificity problems with phenotypic tests [89]. However, molecular methods require expensive equipment and trained laboratory stuff.

6. Therapeutic options

Colistin is usually the last resort antibiotic for the treatment of infections associated with carbapenemase-producing Gram-negative bacteria. However, nephrotoxicity and neurotoxicity are of clinical concern [94]. Colistin-resistant isolates of *A. baumannii*, *P. aeruginosa*, and *K. pneumoniae* have emerged recently leading to pan-drug resistant phenotypes with modified outer membrane lipopolysaccharide [95–97]. The optimal treatment approach for infection caused by carbapenemase-producing Gram-negative bacilli remains a controversy. Still, a combination therapy is strongly recommended and includes two or more antibiotics active in vitro such as colistin, tigecycline, amikacin, aztreonam, or carbapenem in different combination schemes [98, 99]. Colistin is not recommended as monotherapy because of the development of heteroresistance. The combinations of colistin with rifampicin, vancomycin, and meropenem were shown to be synergistic in *A. baumannii* in vitro by chequerboard, 2-well and time-kill metod, but the results of the randomized, controlled clinical trials are controversial [99–101]. If the pathogen is suspected to be MBL producer, aztreonam may present the core drug (if the isolate does not possess other extended-spectrum or ampC β-lactamase), but if the patient is critically ill, the combination therapy with three drugs is preferred. If aminoglycosides or fosfomycin retain in vitro activity, for any serious infection, then they should not be using monotherapy. However, oral fosfomycin should not be used for the management of any infection outside the urinary tract, and in critically ill patients, combination therapy should be considered. Tigecycline is recommended for skin and soft-tissue infection with carbapenemase-producing *A. baumannii* and *Enterobacteriacea*e. If MICs to meropenem are low or moderately high (up to 16 μg/ml), high-dose meropenem can be administered by prolonged infusion [99, 100]. Dual carbapenem-based regimen for carbapenem-resistant *Enterobacteriaceae* usually includes high-dose meropenem or high-dose doripenem with ertapenem, and it is also most effective if the third drug is added [99]. In bloodstream and respiratory tract infections colistin could be added, while in gastrointestinal and biliary tract infections colistin and tigecycline should be included. Newer β-lactam-β-lactamase inhibitor combinations such as ceftolozan/tazobactam seem promising in treatment of carbapenem-resistant *Pseudomonas aeruginosa* infections [102]. However, it is not effective against MBL-producing organisms. Continued reports should be acknowledged in efforts of optimizing therapy for infections caused by carbapenem-resistant Gram-negative bacili. Furthermore, there is an urgent need to develop new antimicrobials. Although antibiotics still represent the mainstream direction for the treatment, the alternatives to them should be taken into account, including vaccines, bacteriocins, and probiotics. The revival of old antibiotics should be supported by expanding the knowledge of their pharmacokinetics and pharmacodynamics in order to avoid side effects and resistance developments.

7. Conclusions

In the last decade, an epidemic spread of carbapenamases among *Enterobacteriaceae* and Gram-negative nonfermentative bacilli was observed worldwide. The type of the most prevalent carbapenemase in a geographic region differs and might be associated with historical or cultural connections and exchange of people between countries where certain types of carbapenemase are endemic including the transfer of patients or staff across the borders, medical tourism, and migration of refugees. Particularly alarming is the fact that carbapenemases are not restricted only to hospital isolates any more. They have been reported and are continuously circulating between hospitals, long-term care facilities, community, and the environment. In conclusion, facing the global dissemination of carbapenemase-producing Gram-negative bacteria, and awaiting for the new antimicrobials to appear, a reasonable approach including hand hygiene, contact precautions, rational antibiotic usage, and active surveillance should be part of an intensive strategy aimed to reduce the incidence of colonization and infections with carbapenem-resistant microorganisms and to stop the spread of these bacteria into the community or the environment and vice versa.

Author details

Branka Bedenić[1,2]* and Sanda Sardelić[3]

1 Department of Microbiology, School of Medicine, University of Zagreb, Zagreb, Croatia

2 Clinical Department of Clinical and Molecular Microbiology, University Hospital Centre Zagreb, Zagreb, Croatia

3 Clinical Department of Microbiology and Parasitology, University Hospital Centre Split, Split, Croatia

*Address all correspondence to: branka.bedenic@kbc-zagreb.hr

References

[1] El-Gamal MI, Brahim I, Hisham N, Aladdin R, Mohammed H, Bahaaeldin A. Recent updates of carbapenem antibiotics. European Journal of Medicinal Chemistry. 2017;**131**:185-195

[2] Queenan AM, Bush K. Carbapenemases: The versatile β-lactamases. Clinical Microbiology Reviews. 2007;**20**(3):440-458

[3] Nordmann P, Poirel L. Emerging carbapenemases in gram-negative aerobes. Clinical Microbiology and Infection. 2002;**8**:321-331

[4] Meletis G. Carbapenem resistance: Overview of the problem and future perspectives. Therapeutic Advances in Infectious Disease. 2016;**3**(1):15-21

[5] Diene SM, Rolain J-M. Carbapenemases genes and genetic platforms in gram-negative bacilli: Enterobacteriaceae, pseudomonas and Acinetobacter species. Clinical Microbiology and Infection. 2014;**20**:831-838

[6] Bush K, Jacoby GA, Medeiros AA. A functional classification scheme for β-lactamases and its correlation with molecular structure. Antimicrobial Agents and Chemotherapy. 1995;**39**:1211-1233

[7] Bush K, Jacoby GA. Updated functional classification of β-lactamases. Antimicrobial Agents and Chemotherapy. 2010;**54**:969-976

[8] Naas T, Vandel W, Sougakoff W, Livermore DM, Nordmann P. Cloning and sequence analysis of the gene for carbapenem hydrolyzing class a β-lactamase, Sme-1 from *Serratia marcescens* S6. Antimicrobial Agents and Chemotherapy. 1994;**38**:1262-1270

[9] Gales AC, Biedenbach DJ, Winokur P, Pfaller A, Jones RN. Carbapenem-resistant *Serratia marcescens* isolates producing group 2F β-lactamase (SME-2) in the unites states: Results from the mystic program. Diagnostic Microbiology and Infectious Disease. 2001;**39**:125-127

[10] Queenan AM, Shang W, Schreckenberg P, Lolans K, Bush K, Quinn J. SME-3 a novel member of the *Serratia marcescens* SME family of carbapenem-hydrolyzing β-lactamases. Antimicrobial Agents and Chemotherapy. 2006;**50**:3485-3487

[11] Nordmann P, Mariotte s NT, Labia R, Nicolas MH. Biochemical properties of a carbapenem-hydrolyzing β-lactamase of *Enterobacter cloacae* and cloning of the gene into *Escherichia coli*. Antimicrobial Agents and Chemotherapy. 1993;**37**:939-946

[12] Prottumarthy S, Moland ES, Jeretschko S, Swanzy R, Thomson KS, Fritsche TR. NMC-A carbapenem-hydrolyzing enyzme in *Enterobacter cloacae* in North America. Emerging Infectious Diseases. 2003;**9**:999-1002

[13] Yigit H, Quennan AM, Anderson GJ, et al. Novel carbapenem-hydrolyzing β-lactamase, KPC-1, from a carbapenem-resistant strain of *Klebsiella pneumoniae*. Antimicrobial Agents and Chemotherapy. 2001;**45**(4):1151-1161

[14] Woodford NP, Tierno PM, Young K, Tysal M, Palepou E, Ward R, Painter E, et al. Outbreak of *Klebsiella pnemoniae* producing a new carbapenem hydrolyzing class a β-lactamase KPC-3 in a New York medical center. Antimicrobial Agents and Chemotherapy. 2004;**48**:4793-4799

[15] Yigit H, Quennan AM, Anderson GJ, et al. Author's correction. Novel carbapenem-hydrolyzing β-lactamase, KPC-1, from a carbapenem-resistant strain of *Klebsiella pneumoniae*.

Antimicrobial Agents and Chemotherapy. 2008;**52**:809

[16] Poirel L, Weldhagen GF, Naas T, De Champs C, Dove G, Nordmann P. GES-2 a novel class a β-lactamase with increased hydrolysis of imipenem. Antimicrobial Agents and Chemotherapy. 2001;**45**:2598-2603

[17] Naas T, Dortet L, Iorga BI. Structural and functional aspects of class A carbapenemases. Current Drug Targets. 2016;**17**(9):1006-1028

[18] Mathers AJ, Stoesser N, Chai W, et al. Chromosomal integration of the *Klebsiella pneumoniae* carbapenemase gene bla_{KPC} in *Klebsiella* species is elusive but not rare. Antimicrobial Agents and Chemotherapy. 2017;**61**(3):1-11. pii: e01823-16. DOI: 10.1128/AAC.01823-16

[19] Nordmann P, Cuzon G, Naas T. The real threat of *Klebsiella pneumoniae* carbapenemase-producing bacteria. The Lancet Infectious Diseases. 2009;**9**(4):228-236

[20] Parisi SG, Bartolini A, Santacatterina E, et al. Prevalence of *Klebsiella pneumoniae* strains producing carbapenemases and increase of resistance to colistin in an Italian teaching hospital from January 2012 to December 2014. BMC Infectious Diseases. 2015;**15**:244. DOI: 10.1186/s12879-015-0996-7

[21] Schwaber MJ, Lev B, Israeli A, Solter E, Smollan G, Rubinovitch B, Shalit I, Carmeli Y, Israel Carbapenem-Resistant Enterobacteriaceae Working Group. Containment of a country-wide outbreak of carbapenem-resistant *Klebsiella pneumoniae* in Israeli hospitals via a nationally implemented intervention. Clinical Infectious Diseases. 2011;**52**(7):848-855. DOI: 10.1093/cid/cir025

[22] Wei DD, Wan LG, Deng Q, Liu Y. Emergence of KPC-producing *Klebsiella pneumoniae* hypervirulent clone of capsular serotype K1 that belongs to sequence type 11 in Mainland China. Diagnostic Microbiology and Infectious Disease. 2016;**85**(2):192-194. DOI: 10.1016/j.diagmicrobio.2015.03.012

[23] Van Duin D, Doi Y. The global epidemiology of carbapenemase-producing Enterobacteriacea. Virulence. 2017;**8**:460-469

[24] Pitout JD, Nordmann P, Poirel L. Carbapenemase-producing *Klebsiella pneumoniae*, a key pathogen set for global nosocomial dominance. Antimicrobial Agents and Chemotherapy. 2015;**59**(10):5873-5884

[25] Poirel L, Thomas IL, Naas T, Karim A, Nordman P. Biochemical sequence analysis of GES-1, a novel class a extended-spectrum β-lactamase and the class I integron In52 from *Klebsiella pneumoniae*. Antimicrobial Agents and Chemotherapy. 2000;**44**:622-632

[26] Poirel L, Weldhagen GF, De Champs C, Nordmann P. A nosocomial outbreak of *Pseudomonas aeruginosa* isolates expressing the extended-spectrum beta-lactamase GES-2 in South Africa. The Journal of Antimicrobial Chemotherapy. 2002;**49**(3):561-565

[27] Walsh T, Toleman MA, Poirel L, Nordmann P. Metallo-β-lactamases: The quiet before the storm. Clinical Microbiology Reviews. 2005;**18**(2):306-325

[28] Berglund F, Marathe NP, Österlund T, Bengtsson-Palme J, Kotsakis S, Flach CF, Larsson DGJ, Kristiansson E. Identification of 76 novel B1 metallo-β-lactamases through large-scale screening of genomic and metagenomic data. Microbiome. 2018;**5**(1):134. DOI: 10.1186/s40168-017-0353-8

[29] Watanabe M, Iyobe S, Inoue M, Mitsuhashi S. Transferable

imipenem resistance in *Pseudomonas aeruginosa*. Antimicrobial Agents and Chemotherapy. 1991;**35**:147-151

[30] Arakawa Y, Murakami M, Suzuki K, et al. A novel integron like-element carrying the metallo-β-lactamase gene bla_{IMP}. Antimicrobial Agents and Chemotherapy. 1995;**39**:1612-1615

[31] Riccio CG, ML MA, Lauretti L, Fontana R, Rossolini GM. Appearance of IMP-1 metallo-β-lactamase in Europe. Lancet. 1999;**353**:899-900

[32] Riccio ML, Franceschini N, Boschi L, et al. Characterization of the metallo-β-lactamase determinant of *Acinetobacter baumannii* AC-54/97 reveals the existance of bla_{IMP} allelic variant carried by gene cassettes of different phylogeny. Antimicrobial Agents and Chemotherapy; **44**:1229-1235

[33] Matsumura Y, Peirano G, Motyl MR, Adams MD, et al. Global molecular epidemiology of IMP-producing enterobacteriaceae. Antimicrobial Agents and Chemotherapy. 2017;**61**(4). DOI: 10.1128/AAC.02729-16

[34] Iyobe S, Kusadokoro H, Ozaki J, Matsumura N, Minami S, Haruta S, Sawai T, O'Hara K. Amino acid substitutions in a variant of IMP-1 metallo-β-lactamase. Antimicrobial Agents and Chemotherapy. 2000;**44**(8):2023-2027

[35] Chu YW, Afzal Shah M, Huang E, Palepou MF, Lyon DJ, Woodford N, Livermore DM. IMP-4, a novel metallo-β-lactamase from nosocomial *Acinetobacter* spp. collected in Hong-Kong between 1994 and 1998. Antimicrobial Agents and Chemotherapy. 2001;**45**(3):710-714

[36] Da Silva GJ, Correia M, Vital C, Ribero G, Sousa JC, Leitao R, Peixe L, Duarte A. Molecular characterization of $bla_{(IMP\text{-}5)}$ a new integron born metallo-β-lactamase gene from an *Acinetobacter baumannii* nosocomial isolate in Portugal. FEMS Microbiology Letters. 2002;**24**(215):33-39

[37] Available from: ftp://ftp.ncbi.nlm.nih.gov/pathogen/βlactamases/Allele.tab [Accessed: 2018-03-06]

[38] Mojica MF, Bonomo RA, Fast W. B1-metallo-β-lactamases: Where do we stand? Current Drug Targets. 2016;**17**(9):1029-1050

[39] Tada T, Nhung PH, Miyoshi-Akiyama T, Shimada K, Phuong DM, Anh NQ, Ohmagari N, Kirikae T. IMP-51, a novel IMP-type metallo-β-lactamase with increased doripenem and meropenem hydrolyzing activities, in a carbapenem-resistant *Pseudomonas aeruginosa* clinical isolate. Antimicrobial Agents and Chemotherapy. 2015;**59**:7090-7093

[40] Cornaglia G, Giamarellou H, Rossolini GM. Metallo-β-lactamases: A last frontier for β-lactams? The Lancet Infectious Diseases. 2011;**11**:381-393

[41] Carattoli A. Resistance plasmid families in Enterobacteriaceae. Antimicrobial Agents and Chemotherapy. 2009;**53**(6):2227-2238. DOI: 10.1128/AAC.01707-08

[42] Lauretti L, Riccio ML, Mazzariol A, Cornaglia G, Amicosante G, Fontana R, Rossolini GM. Cloning and characterization of bla_{VIM}, a new integron-borne metallo- β-lactamase gene from *Pseudomonas aeruginosa* clinical isolate. Antimicrobial Agents and Chemotherapy. 1999;**43**(7):1584-1590

[43] Poirel L, Naas T, Nicolas D, Collet L, Bellais S, Cavallo JD, Nordmann P. Characterization of VIM-2, a carbapenem-hydrolyzing metallo-β-lactamase and its plasmid and integron-borne gene from

Pseudomonas aeruginosa clinical isolate in France. Antimicrobial Agents and Chemotherapy. 2000;**44**:891-897

[44] Yan JH, Po-Ren H, Wen-Chien K, et al. Metallo-β-lactamases in clinical *Pseudomonas* isolates in Taiwan and identification of VIM-3, a novel variant of the VIM-2 enzyme. Antimicrobial Agents and Chemotherapy. 2001;**458**:2224-2228

[45] Luzzaro F, Docquier JD, Colinon C, et al. Emergence in *Klebsiella pneumoniae* and *Enterobacter cloacae* clinical isolates of the VIM-4 metallo β-lactamase encoded by a conjugative plasmid. Antimicrobial Agents and Chemotherapy. 2004;**48**:648-650

[46] Zhao W-H, Hu Z-Q. Epidemiology and genetics of VIM-type metallo-β-lactamase in Gram-negative bacilli. Future Microbiology. 2011;**6**:317-333

[47] Toleman MA, Rolson K, Jones RN, Walsh T. Bla_{VIM-7} and evolutionary distinct metallo-β-lactamase gene in a *Pseudomonas aeruginosa* isolate from the United States. Antimicrobial Agents and Chemotherapy. 2004;**48**:329-332

[48] Samuelsen O, Castanheira M, Walsh TR, Spencer J. Kinetic characterization of VIM-7, a divergent member of the VIM metallo-beta-lactamase family. Antimicrobial Agents and Chemotherapy. 2008;**52**(8):2905-2908. DOI: 10.1128/AAC.00166-08

[49] Kumar V, Sen MR, Nigam C, Gahlot R, Kumari S. Burden of different beta-lactamase classes among clinical isolates of AmpC-producing *Pseudomonas aeruginosa* in burn patients: A prospective study. Indian Journal of Critical Care Medicine. 2012;**16**(3):136-140

[50] Oelschlaeger P, Ai N, Duprez KT, Welsh WJ, Toney JH. Evolving carbapenemases: Can medicinal chemists advance one step ahead of the coming storm? Journal of Medicinal Chemistry. 2010;**53**:3013-3027

[51] Yong D, Toleman MA, Giske CG, Cho HS, Sundman K, Lee K, Walsh TR. Characterization of a new metallo-β-lactamase gene, bla_{NDM-1}, and a novel erythromycin esterase gene carried on a unique genetic structure in *Klebsiella pneumoniae* sequence type 14 from India. Antimicrobial Agents and Chemotherapy. 2009;**53**(12):5046-5054

[52] Walsh TR, Weeks J, Livermore DM, Toleman MA. Dissemination of NDM-1 positive bacteria in New Delhi environment and its implications for human health: An environmental point prevalence study. The Lancet Infectious Diseases. 2011;**11**:355-362

[53] Lee C-R, Lee JH, Park KS, Kim YB, Jeong BC, Lee SH. Global dissemination of carbapenemase producing *Klebsiella pneumoniae*: Epidemiology, genetic context, treatment options and detection methods. Frontiers in Microbiology. 2016;**7**:895. DOI: 10.3389/frmicb.2016.00895

[54] Khan AU, Maryam L, Zarrilli R. Structure, genetics and worldwide spread of New Delhi metallo-β-lactamase (NDM): A treat to public health. BMC Microbiology. 2017;**17**:101. DOI: 10.1186/s12866-017-1012-8

[55] Kashyap A, Gupta R, Sharma R, Verma VV, Gupta S, Pradeep G. New Delhi metallo beta lactamase: Menace and its challenges. Journal of Molecular and Genetic Medicine. 2017;**11**:4. DOI: 10.4172/1747-0862.1000299

[56] Kazmiercak KM, Rabine S, Hackel M, McLaughlin RE, Biedenbach DJ, Bouchillon SK, Sahm DF, Bradford PA. Multiyear, multinational survey of the incidence and global distribution of metallo-β-lactamase-producing Enterobacteriaceae and

Pseudomonas aeruginosa. Antimicrobial Agents and Chemotherapy. 2015;**60**(2):1067-1078. DOI: 10.1128/AAC.02379-15

[57] Toleman MA, Simm AM, Murphy TA, Gales AC, Biedenbach DJ, Jones RN, Walsh TR. Molecular characterization of SPM-1, a novel metallo β-lactamase isolated in Latin America: Report from the SENT antimicrobial surveillance programme. The Journal of Antimicrobial Chemotherapy. 2002;**50**:673-679

[58] Castanheira M, Toleman MA, Jones RN, Schmidt FJ, Walsh TR. Molecular characterization of β-lactamase gene, bla_{GIM} encoding a new subclass of metallo β-lactamase. Antimicrobial Agents and Chemotherapy. 2004;**48**:4654-4661

[59] Lee K, Yum JH, Yong D, Chong J, Kim JM, Docquirer JD, Rossolini GM, Chong Y. Novel acquired metallo-β-lactamase gene, bla_{SIM-1}, in a class 1 integron from *Acinetobacter baumannii* clinical isolates, from Korea. Antimicrobial Agents and Chemotherapy. 2005;**49**:4485-4491

[60] Deshpande LM, Jones RN, Woosley LN, Castanheira M. Retrospective molecular analysis of DIM-1 metallo-β-lactamase discovered in *Pseudomonas stutzeri* from India in 2000. Antimicrobial Agents and Chemotherapy. 2014;**58**:596-598

[61] Salabi AE, Toleman MA, Weeks J, Bruderer T, Frei R, Walsh TR. First report of the metallo-beta-lactamase SPM-1 in Europe. Antimicrobial Agents and Chemotherapy. 2010;**54**:582

[62] Hopkins KL, Meunier D, Findlay J, Mustafa N, Parsons H, Pike R, Wright L, Woodford N. SPM-1 metallo-β-lactamase-producing *Pseudomonas aeruginosa* ST277 in the UK. Journal of Medical Microbiology. 2016;**65**(7):696-697. DOI: 10.1099/jmm.0.000269

[63] Shahcheraghi F, Abbasalipour M, Feizabadi M, Ebrahimipour G, Akbari N. Isolation and genetic characterization of metallo-β-lactamase and carbapenamase producing strains of *Acinetobacter baumannii* from patients at Tehran hospitals. Iranian Journal of Microbiology. 2011;**3**:68-74

[64] Brown S, Amyes S. OXA β-lactamase in *Acinetobacter*: The story so far. The Journal of Antimicrobial Chemotherapy. 2006;**57**:1-3

[65] Evans BA, Amyes SG. OXA β-lactamases. Clinical Microbiology Reviews. 2014;**27**:241-263. DOI: 10.1128/CMR.00117-13

[66] Da Silva GJ, Domingues S. Insights on the horizontal gene transfer of carbapenemase determinants in the opportunistic pathogen *Acinetobacter baumannii*. Microorganisms 2016;4:29. doi:10.3390/microorganism4030029

[67] Paton RH, Miles RS, Hood J, Amyes SGB. ARI-1: β-lactamase mediated imipenem resistance in *Acinetobacter baumannii*. International Journal of Antimicrobial Agents. 1993;**2**:81-88

[68] Bou G, Oliver A, Martinez-Beltran J. OXA-24, a novel class D β-lactamase with carbapenemase activity in an *Acinetobacter baumannii* clinical strain. Antimicrobial Agents and Chemotherapy. 2000;**44**:1556-1561

[69] Marque S, Poirel L, Heritier C, et al. Regional occurrence of plasmid-mediated carbapenem-hydrolyzing oxacillinase OXA-58 in *Acinetobacter* spp. in Europe. Journal of Clinical Microbiology. 2005;**43**:4885-4888

[70] Higgins P, Poirel L, Lehmann M, Nordmann P, Seifert H. OXA-143, a novel carbapenem-hydrolyzing class D β-lactamase in *Acinetobacter baumannii*. Antimicrobial Agents and Chemotherapy. 2009;**53**:5035-5038

[71] Goić-Barišić I, Towner KJ, Kovačić A, Šiško-Kraljević K, Tonkić M, Novak A, Punda-Polić V. Outbreak in Croatia caused by a new carbapenem-resistant clone of *Acinetobacter baumannii* producing OXA-72 carbapenemase. Journal of Hospital Infection. 2011;**77**:368-369

[72] Wang H, Guo P, Sun H, Wang H, Yang Q, Chen M, Xu Y, Zhu Y. Molecular epidemiology of clinical isolates of carbapenem-resistant *Acinetobacter* spp. from Chinese hospitals. Antimicrobial Agents and Chemotherapy. 2007;**51**: 4022-4028

[73] Werneck JS, Picao RC, Carvalhaes CG, Cardoso JP, Gales A. OXA-72-producing *Acinetobacter baumannii* in Brazil: A case report. The Journal of Antimicrobial Chemotherapy. 2011;**66**(2):452-454

[74] Bedenić B, Beader N, Godič-Torkar K, Vranić-Ladavac M, Luxner J, Veir Z, Grisold AJ, Zarfel G. Nursing home as reservoir of carbapenem-resistant *Acinetobacter baumannii*. Microbial Drug Resistance. 2015;**21**(3):270-278

[75] Hrenović J, Goić-Barišić I, Kazazic S, Kovačić A, Ganjto M, Tonkić M. Carbapenem-resistant isolates of *Acinetobacter baumannii* in a municipal wastewater treatment plant, Croatia, 2014. Euro Surveillance. 2016;**21**(15):pii=30195. DOI: http://dx.doi.org/10.2807/1560-7917.ES.2016.21.15.30195

[76] Goić-Barišić I, Hrenović J, Kovačić A, Šeruga-Musić M. Emergence of oxacillinases in environmental carbapenem-resistant *Acinetobacter baumannii* associated with clinical isolates. Microbial Drug Resistance. 2016;**22**(7):559-563

[77] Poirel L, Heritier C, Tolun V, Nordmann P. Emergence of oxacillinases-mediated resistance to imipenem in *Klebsiella pneumoniae*. Antimicrobial Agents and Chemotherapy. 2004;**48**(1):15-22

[78] Poirel L, Potron A, Nordmann P. OXA-48-like carbapenemases: The phantom menace. Journal of Antimicrobial Chemotherapy. 2012;**67**(7):1597-1606. DOI: 10.1093/jac/dks121

[79] Gulmez D, Woodford N, Palepou MF, et al. Carbapenem resistant *Escherichia coli* and *Klebsiella pneumoniae* isolates from Turkey with OXA-48 like carbapenemase and outer membrane protein loss. International Journal of Antimicrobial Agents. 2008;**31**(6):523

[80] Poirel L, Ros A, Carricajo A, et al. Extremely drug-resistant *Citrobacter freundii* isolate producing NDM-1 and other carbapenemases identified in a patient returning from India. Antimicrobial Agents and Chemotherapy. 2011;**55**:447-448

[81] Potron A, Poirel L, Nordmann P. Origin of OXA-181, an emerging carbapenem-hydrolyzing oxacillinase, as a chromosomal gene in *Shewanella xiamenensis*. Antimicrobial Agents and Chemotherapy. 2011;**55**:4405-4407

[82] Borer A, Saidel-Odes L, Riesenberg K, Eskira S, Peled N, Nativ R, Schlaeffer F, Sherf M. Attributable mortality rate for carbapenem-resistant *Klebsiella pneumoniae* bacteremia. Infection Control and Hospital Epidemiology. 2009;**30**:972-976

[83] Zilberberg MD, Nathanson BH, Sulham K, Fan W, Shorr AF. Carbapenem resistance, inappropriate empiric treatment and outcomes among patients hospitalized with *Enterobacteriaceae* urinary tract infection, pneumonia and sepsis. BMC Infectious Diseases. 2017;**17**:279. DOI: 10.1186/s12879-017-2383-z

[84] Akova DGL, Tzouvelekis L, Carmeli Y. Interventional strategies and current clinical experience with carbapenemase-producing gram-negative bacteria. Clinical Microbiology and Infection. 2012;**18**:439-448

[85] Clinical and Laboratory Standards Institute. Performance Standards for Antimicrobial Susceptibility Testing. 27th ed. CLSI Document M100S. Wayne, PA, USA: CLSI; 2017

[86] The European Committee on Antimicrobial Susceptibility Testing. Breakpoint Tables for Interpretation of MICs and Zone Diameters. Version 7.1. 2017. Available online: http://www.eucast.org [Accessed: 2018-03-06]

[87] Miriagou V, Cornaglia G, Edelstein M, et al. Acquired carbapenemases in gram-negative bacterial pathogens: Detection and surveillance issues. Clinical Microbiology and Infection. 2010;**16**:112-122

[88] Nordmann P, Gniadkowski M, Giske CG, Poirel L, Woodford N, Miriagou V, European Network on Carbapenemases. Identification and screening of carbapenemase-producing enterobacteriaceae. Clinical Microbiology and Infection. 2012;**18**:432-438

[89] Madkour LA, Soliman MS, Hassan DM, Soliman NS, ElMahdy YA. Detection of carbapenemase-producers: Evaluating the performance of the carbapenem inactivation method and Carba NP test versus multiplex PCR. Journal of Global Antimicrobial Resistance. 2017;**9**:10-14

[90] Aguirre-Quiñonero A, Cano ME, Gamal D, Calvo J, Martínez-Martínez L. Evaluation of the carbapenem inactivation method (CIM) for detecting carbapenemase activity in enterobacteria. Diagnostic Microbiology and Infectious Disease. 2017;**88**:214-218

[91] Sun K, Xu X, Yan J, Zhang L. Evaluation of six phenotypic methods for the detection of carbapenemases in gram-negative bacteria with characterized resistance mechanisms. Annals of Laboratory Medicine. 2017;**37**(4):305-312

[92] Pierce VM, Simner PJ, Lonsway DR, et al. Modified carbapenem inactivation method for phenotypic detection of carbapenemase production among enterobacteriaceae. Journal of Clinical Microbiology. 2017;**55**(8):2321-2333

[93] Johansson A, Ekelöf JJ, Giske CG, Sundqvist M. The detection and verification of carbapenemases using ertapenem and matrix assisted laser desorption ionization-time of flight. BMC Microbiology. 2014;**14**:89. DOI: 10.1186/1471-2180-14-89

[94] Michalopoulos A, Falagas ME. Colistin and polymyxin B in critical care. Critical Care Clinics. 2008;**466**(24):377-391

[95] Cai Y, Chai D, Wang R, Liang B, 638 Bai N. Colistin resistance of *Acinetobacter baumannii*: Clinical reports, mechanisms and antimicrobial strategies. The Journal of Antimicrobial Chemotherapy. 2012;**67**:1607-1615

[96] Ko KS, Suh JY, Kwon KT, Jung S-I, et al. High rates of resistance to colistin and polymyxin B in subgroups of *Acinetobacter baumannii* isolates from Korea. The Journal of Antimicrobial Chemotherapy. 2007;**60**:1163-1167

[97] Helander IM, Kato Y, Kilpeläinen I, Kostiainen R, Lindner B, Nummila K, et al. Characterization of lipopolysaccharides of polymyxin-resistant and polymyxin-sensitive *Klebsiella pneumoniae*. European Journal of Biochemistry. 1996;**23**:272-278

[98] Paul M, Carmeli Y, Durante-Mangoni E, et al. Combination therapy for carbapenem-resistant gram-negative

bacteria. The Journal of Antimicrobial Chemotherapy. 2014;**69**(9):2305

[99] Trecarichi EM, Tumbarello M. Therapeutic options for carbapenem-resistant *Enterobacteriaceae* infections. Virulence. 2017;**8**:470-484

[100] Gilad J, Carmely Y. Treatment options for multidrug-resistant *Acinetobacter* spp. Drugs. 2008;**68**:165-189

[101] Petrosillo N, Ionnidou E, Falagas ME. Colistin monotherapy vs. combination therapy: Evidence from microbiological, animal and clinical studies. Clinical Microbiology and Infection. 2008;**14**:816-827

[102] Munita JM, Aitken SL, Miller WR, et al. Multicenter evaluation of ceftolozane/tazobactam for serious infections caused by carbapenem-resistant *Pseudomonas aeruginosa*. Clinical Infectious Diseases. 2017;**65**:158-161

Chapter 4

Design and Operation of Fixed-Bed Bioreactors for Immobilized Bacterial Culture

Ralf Pörtner and Rebecca Faschian

Abstract

Fixed-bed processes operated in perfusion, where cells are immobilized within macroporous carriers, are a promising alternative to processes with suspended microbial or mammalian cells. Their potential has been demonstrated for many purposes. Nevertheless, the number of industrial fixed-bed processes is quite small. To some extent, this is due to the lack of process development tools for fixed-bed processes. To fill this gap, a strategy was developed for the design and evaluation of relevant process parameters of fixed-bed processes. A scale-up concept is presented in order to evaluate the performance as part of process design of fixed-bed processes. This comprises fixed-bed reactors on three different scales, the smallest being the downscaled Multiferm with 10 mL fixed-bed units, the second a 100 mL fixed-bed reactor, and the third a pilot-scale reactor with 1 L fixed-bed volume. The performance of this concept will be discussed for fixed-bed cultures of lactic acid bacteria. Furthermore, a reaction kinetic model for the design of fixed-bed reactors will be presented.

Keywords: fixed-bed, bacterial culture, scale-up, modeling, lactic acid bacteria

1. Introduction

Technologies for immobilization of biocatalyst, e.g., microbial or mammalian cells, are increasingly being considered for biotechnological processes due to many advantages compared to cell suspension culture such as continuous operation, accelerated reaction rates, high volumetric productivity, retention of plasmid-bearing cells, prevention of interfacial inactivation, stimulation of production and excretion of secondary metabolites, and protection against turbulent high-shear environment, reduced susceptibility of cells to contamination, improved production efficiency, and reduced risk of washout [1–3]. Especially the increased importance of productivity for industrial processes due to restriction of production time and final product volume has drawn the attention to immobilization techniques in recent years, as they allow overcoming most of the limitations of commonly applied suspension cultures [2]. A summary of advantages and disadvantages for suspension cultures (stirred tank reactors) and immobilized cultures (fixed-bed reactors) is given in **Table 1**.

Various immobilization techniques such as the entrapment of cells in stable porous gels (e.g., alginate, agarose, collagen, chitosan, cellulose, κ-carrageenan, or

	Advantages	Disadvantages
Stirred tank/ suspension	Known technology	Aeration difficult at high cell densities (relevant for aerobic cells)
	Good mass transfer	Cell damage by shear and aeration (e.g., mammalian cells)
	Good mixing	Foaming (relevant for aerobic cells)
	Cell count possible	Low cell density and volumetric productivity Cell retention required for perfusion culture, techniques insufficient for long-term culture
	High potential for scale-up	
Fixed-bed/ immobilized cells	High cell density and productivity per unit	Concentration gradients
	Easy exchange of medium	Nonhomogeneous
	High productivity over long periods of time	Cell count impossible
	Low-shear rates (relevant for mammalian cells)	

Table 1.
Summary of advantages and disadvantages of stirred tank (suspended cells) and fixed-bed reactors (immobilized cells).

- High volumetric cell density and high productivity
- Low-shear stress environment for mammalian and tissue cells
- Easy medium exchange and separation of cells and product simplifying downstream processing

Table 2.
Characteristics of fixed-bed reactors used for cultivation of microorganisms or mammalian/tissue cells (adapted from Pörtner and Märkl [5]).

gel-matrix polymers such as polyacrylamide-hydrazide) or hydrogels or immobilization in solid macroporous carriers have been developed and are applied in both laboratory and industrial scales for different purposes, e.g., food, dairy, and beverage industry, production of drugs, wastewater treatment, agricultural industry, and biodiesel production [2–4].

Bioreactors for immobilized biocatalyst are mostly operated continuously in perfusion mode. Here continuous stirred tank reactors with cell retention and fixed-bed (packed bed) or fluidized-bed bioreactor systems can be applied. The following remarks focus on fixed-bed reactors, which consist of a packed column of macroporous carriers wherein cells are immobilized, as they have been used very successfully for a wide range of applications [4]. The advantages of fixed-bed reactors with immobilized cells (**Table 2**) are mainly with respect to general productivity and operational flexibility [6]. The volumetric productivity of immobilized cells is generally higher than the corresponding free cell fermentations [6]. This higher productivity can be explained by the fact that the microenvironments offered by the carrier are more stabilizing for the organisms, which generally show optimal activity only in a narrow range of physical conditions. Due to cell retention, it is possible to run fixed-bed bioreactors in a perfusion mode at a steady state with dilution rates higher than the maximum specific growth rate of the used strain. By this, very high volume-specific productivities can be reached and maintained for long periods of time and greatly facilitate recycling or reuse of microorganisms [6]. Consequently, both the operational stability of the immobilized organisms and the productivity are improved.

Despite the obvious advantages of fixed-bed bioreactor systems, the number of industrial fixed-bed processes is quite small [2–4]. To some extent, this is due to the lack of process development tools for fixed-bed processes and meaningful concepts for design and operation of fixed-bed reactors on a large scale. To fill this gap, strategies for the design and evaluation of relevant process parameters of fixed-bed processes are required, and a scale-up concept is introduced.

In the following, the characteristics of fixed-bed bioreactors as well as a design concept for layout and scale-up will be introduced. Examples for macroporous carriers will be given. The design strategy for fixed-bed reactors will be discussed in detail for immobilized cultures of lactic acid-producing bacteria (LABs). Finally, a reaction kinetic model is introduced which allows evaluation of the culture performance. Conclusions complete the text.

2. Fixed-bed reactor systems

2.1 Principle

Fixed-bed bioreactors consist of a mostly cylindrical column containing macroporous carriers, wherein cells are immobilized (**Figure 1A**). The column is permanently perfused with fresh medium [4, 7]. If required, the medium can be circulated in a loop (**Figure 1B**). This might be useful if appropriate flow rates and medium supply rates vary significantly.

For small fixed-bed volumes with a height of approximately 10 cm, the medium can be pumped axially through the bed. In this case, at the outlet the oxygen concentration in the case of aerobic cells, e.g., mammalian cells [8], or the pH in the case of acid-producing anaerobic cells, e.g., LABs [6], should remain in a physiological range. A further increase of the length would result in too low oxygen or pH values in the upper zones of the bed. This can be overcome by applying a radial medium flow as shown in **Figure 1C**, where the radius determines the length of the oxygen or pH gradient, not the height of the column. This concept was successfully applied for mammalian cell culture [8] and lactic acid bacteria, as discussed in the following.

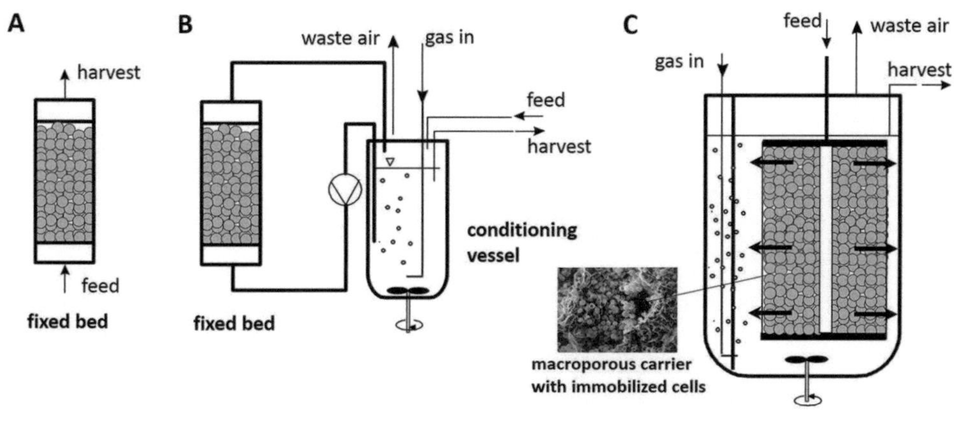

Figure 1.
Examples for design concepts of fixed-bed reactors. (A) Axial-flow fixed bed with plug flow. (B) Axial-flow fixed bed with external conditioning vessel. (C) Fixed bed with radial flow integrated in conditioning vessel, plug flow.

Cell immobilization in fixed-bed reactors with macroporous carriers is fairly simple compared to other methods such as entrapping in gels (e.g., alginate). Cell loading is often carried out by simply pumping a cell suspension through the bed of carriers, and cells are kept under same physiological conditions for the immobilization. As only the natural properties of the surface and cells interact, there are no toxic effects arising from activating reagents compared to cell entrapment within polymers. Additionally, high load of cells can be avoided by desorption of cells from the solid surface to the cell suspension.

2.2 Concept for design and operation

Process parameters that have to be optimized during process development comprise selection of carriers, medium selection, appropriate flow velocity, and long-term performance, among others. All these information are required to evaluate the overall performance, e.g., productivity, and to layout the scale-up strategy. In the following, a platform for development of processes for immobilized cells is introduced (**Figure 2**). As a start, suitability of different carriers can be compared in a small-scale multi-well system. After this, bioreactor systems of different sizes can be used to work out the required process parameters. The first, very small scale of 10 mL working volume is the multi-fixed-bed bioreactor "Multiferm" [9]. The next step is an axial-flow 100 mL fixed-bed system, which can be operated continuously with reasonable effort to investigate the performance and long-term stability of the culture [6]. As a first approach for scale-up, a radial-flow 1 L fixed-bed reactor is applied [6, 10]. Even if this is probably not the final industrial scale, the reactor system has already been the main characteristics of a large-scale system, mainly the radius. For further increase of the volume, just the height has to be increased [11]. For all three systems, a "proof of concept" has been shown before [10]. In Chapter 4 the performance of these three fixed-bed systems is compared for fixed-bed cultures of LABs.

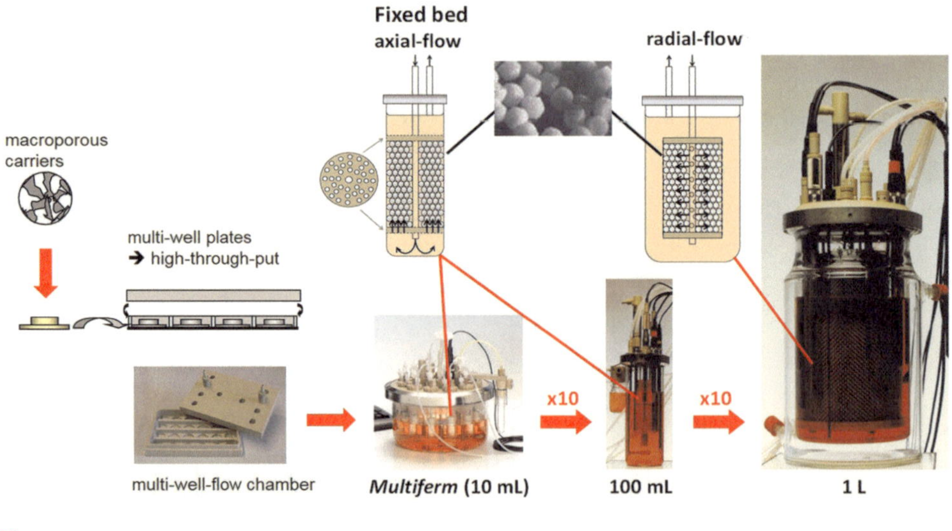

Figure 2.
Platform for development of processes for immobilized cells. From left to right: multi-well plates with special inserts for the first evaluation of appropriate carriers under high-throughput, static conditions; multi-well flow chamber for evaluation of carriers under flow conditions; multi-fixed-bed reactor "Multiferm" (max. 12 small fixed-bed units containing approx. 10 mL carriers for evaluation of carriers under different process parameters (type of carrier, flow rate, medium, oxygen concentration, pH, etc.)); 100 mL axial-flow fixed-bed for long-term continuous culture under steady-state conditions; and 1 L radial-flow fixed-bed representative for a pilot scale for long-term continuous cultures under steady-state conditions.

3. Carriers

In cell immobilization, properties of the carrier materials play an important role. This type of immobilization on solid synthetic materials firstly has the advantage that the microorganisms attach independently to the carrier (interaction with the surface) and thus no additional process steps and reagents are required for immobilization. At this point, carrier materials have to demonstrate several certain characteristics. Atkinson et al. [12] and Pörtner and Märkl [5] summarized these properties for cell immobilization such as simple and nontoxic material, high cell loading capacity, mechanical stability, stable at appropriate operational pH values, autoclavable, resistant to microbial degradation, cost appropriate to the application, density appropriate to reactor type used, as well as reusable, if possible. Examples are given in [4, 6]. In our own studies, carriers made of glass [Siran (QVF, Mainz, Germany), VitraPOR® (ROBU® Glasfilter-Geräte GmbH)] or ceramics [(CERAMTEC EO 19/30 (CeramTec, Marktredwitz, Germany) (**Figure 3**) or Sponceram (Zellwerk, Oberkrämer, Germany)] were applied. All carriers showed similar results with respect to immobilized cell density and lactic acid productivity for immobilized LAB strains [6].

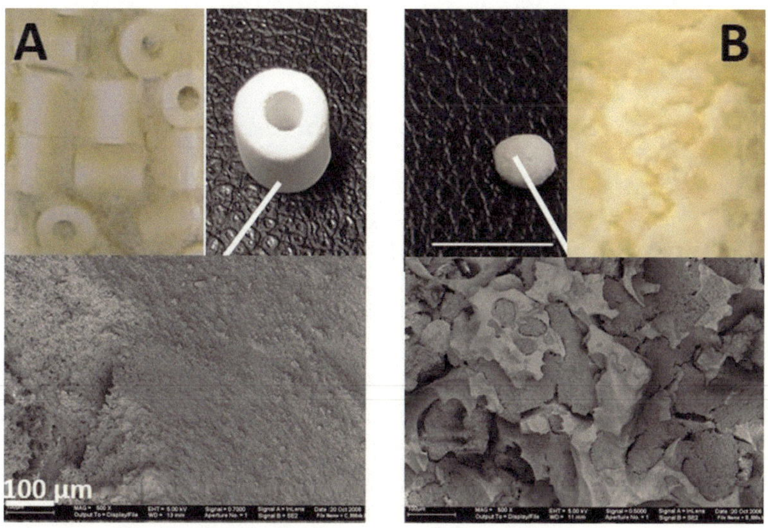

Figure 3.
Examples for carriers applied in fixed-bed cultures: (A) CERAMTEC EO 19/30 (α-aluminum, ring, diameter 3–8 mm, height 8 mm, porosity 65%; manufacturer, CeramTec; up-right, carrier; up-left, carrier in fixed-bed cultivation of Lactococcus lactis*; down, SEM of carrier); (B) VitraPOR® (glass, sphere, diameter 4 mm; manufacturer, ROBU; up-left, carrier; up-right, carrier in fixed-bed cultivation of* L. lactis*; down, SEM of carrier).*

4. Case study: fixed-bed cultivation of LAB strains

4.1 Overview on immobilization techniques used for LAB strains

Lactic acid bacteria are commonly used in the production of fermented dairy products as well as for production of lactic acid, antimicrobial substances (bacteriocins), and biodegradable polymers, among others [13–15]. Industrial processes use mostly conventional batch or fed-batch fermentation with suspended cells. Reactor volumes go up to 100 m^3, and process time varies between several hours and days depending on the strain and the process strategy [14]. Even if high cell and product concentrations can be reached, the known drawbacks such as low productivity, product inhibition, and also the variation from batch to batch remain [16–18].

Since the immobilization of LAB has many advantages, it has been examined extensively, e.g., for the production of lactic acid; the production of starter cultures; the production of bacteriocins, e.g., nisin; and the formation of aromatic compounds (reviewed in [6]). Different methods have been used for immobilizing LAB: physical entrapment in polymeric networks, microencapsulation, attachment or adsorption to a carrier, and membrane entrapment [6]. The purpose of all these techniques is either to keep high cell concentrations within the bioreactor or to protect cells from a hostile environment. In many applications of cell entrapment, droplets of thermal (κ-carrageenan, gellan, agarose, gelatin) or ionotropic (alginate, chitosan) gels are used to produce spherical gel biocatalysts, and these controlled-size polymer droplets are produced using extrusion or emulsification, under mild conditions (reviewed in [6]). However, although promising on a laboratory scale, the large-scale production of beads under aseptic conditions still has difficulties [19].

Another immobilization technique is to immobilize LAB cells onto solid macroporous carriers and apply these in fixed-bed bioreactors. Examples are given in [6]. In the following, our recent work in this area will be discussed.

4.2 Fixed-bed cultures of LAB strains

4.2.1 Examples for fixed-bed cultivation on different scales

For all three fixed-bed bioreactor systems (*Multiferm* 10 mL, axial flow 100 mL, radial flow 1 L), a "proof of concept" has been shown before [6, 9, 10]. More infos on Materials and Methods can be found there. In the following, the main results are highlighted. Both *Lactococcus lactis* subsp. *lactis* and *Lactobacillus delbrueckii* subsp. *bulgaricus* could be cultivated successfully in the fixed-bed reactors. As expected, the lactate concentrations in the harvest flow were in a similar range or slightly lower as in the corresponding batch cultures. The yield of lactate depended on the type of strain and the used medium. The cell concentration in the harvest flow was considerably lower as in the corresponding batch culture, especially in case of *L. bulgaricus*. This is probably due to the short duration of most experiments (50–100 h per perfusion rate). In longer experiments considerably higher cell concentrations in the harvest flow were found.

The volume-specific lactic acid and cell productivity increased with increasing perfusion rate (see below).

Parallel cultivation in the *Multiferm* bioreactor system showed a very high reproducibility [9]. Standard deviation for lactate concentration from different parallel runs was below 5%, indicating a high reproducibility of the system. Therefore, the system is well suited for evaluation of process parameters in a very small scale with reduced effort.

The microbiological and mechanical stabilities of continuous cultivations during prolonged fermentations are critical properties of an immobilized cell process, and industrial applications are largely dependent on these properties. Therefore, we focused on the examination of long-term (52 days) continuous cultivation of *L. lactis* immobilized on ceramic carriers in an axial-flow 100 ml fixed-bed reactor (**Figure 4**) [6]. This proved that the continuous immobilized cell fermentation with *L. lactis* demonstrated a high biological stability longer than 50 days. The viability of cells in the harvest flow was usually around 90%, and the growth rate of cells re-cultivated as batch was similar to the corresponding batch. This indicates that functional cells can be harvested continuously from the fixed-bed.

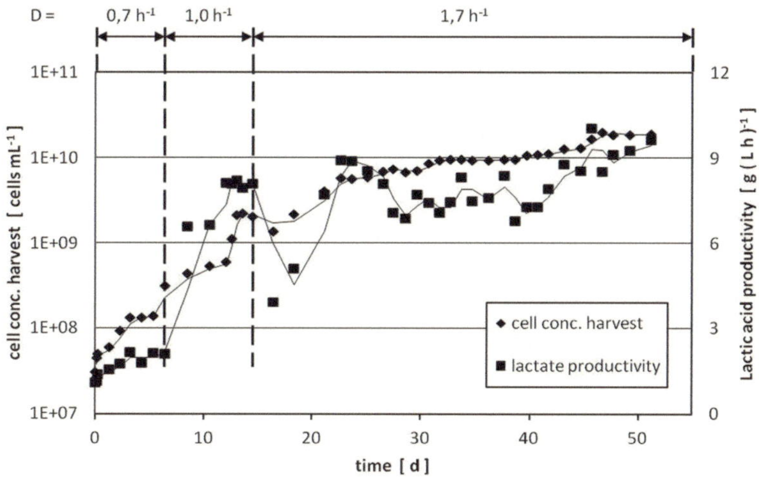

Figure 4.
Long-term cultivation of Lactococcus lactis *in an axial-flow fixed-bed (100 mL, Medorex) filled with macroporous carrier CERAMTEC EO/90 at different perfusion rates D in MRS medium (data from* S. Zengen *(TU Hamburg), not published). Cell concentration in harvest flow (left) and volume-specific lactate productivity versus cultivation time.*

The scale-up from 100 mL to 1 L fixed-bed (**Figure 5**) was successful, as similar productivities could be obtained in both systems (see below). As a conclusion, the continuous cultivation of immobilized LAB strains in fixed-bed reactors shows a high biological stability as well as cell and lactate production in long-term fermentation.

4.2.2 Comparison of suspension and fixed-bed systems on different scales

Fixed-bed and suspension cultures of *L. lactis* were compared with respect to the volume-specific lactate productivity (**Figure 6**) [10]. Continuous suspension culture in chemostat mode showed the expected course [20]. At first the productivity increases with increasing dilution rate up to a maximum. When the dilution rate

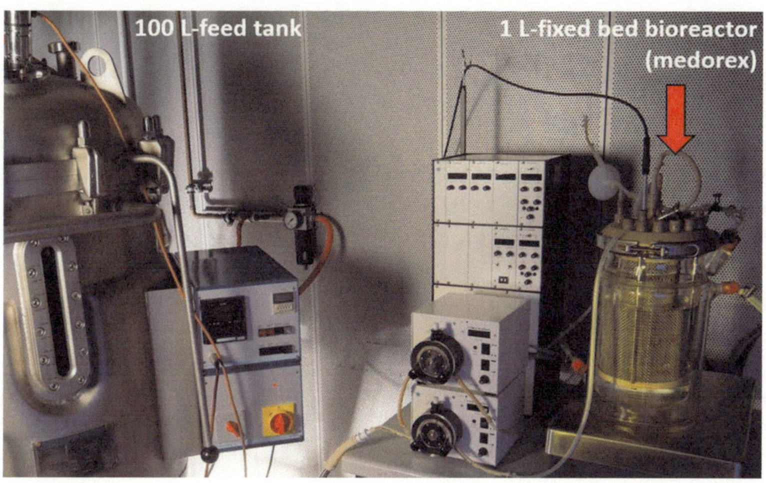

Figure 5.
Pilot plant fixed-bed bioreactor system consisting of 1 L radial-flow fixed-bed reactor (Medorex), feed and harvest pumps, feed tank (100 L), and control unit (temperature, pH, oxygen, depending on the type of microorganism (aerobic or anaerobic)).

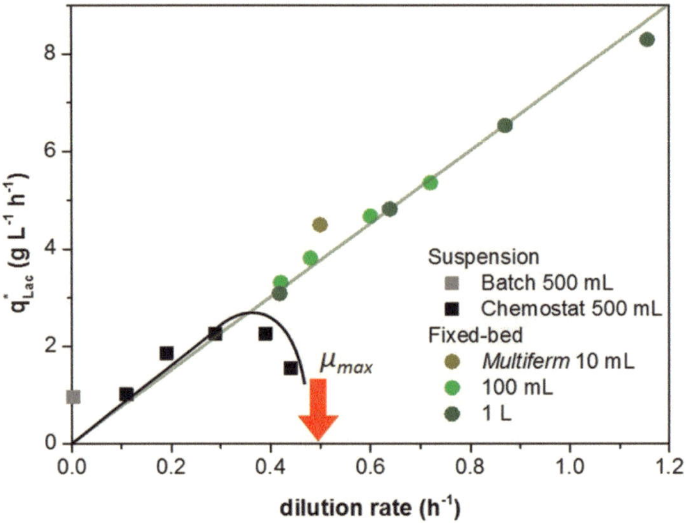

Figure 6.
Comparison of suspension and fixed-bed cultures for Lactococcus lactis. *Volume-specific lactate productivity* q^*_{Lac} *versus dilution rate D. M17 medium (Difco) with 5 g* L^{-1} *lactose; carrier, 10 mL Multiferm; 100 mL, VitraPOR® 4 mm; 1 L, VitraPOR® 8 mm. The red arrow indicates the maximum specific growth rate of the strain.*

gets close to the maximum specific growth rate μ_{max}, the productivity decreases, as washout of cells occurs.

For the fixed-bed-cultures, the productivity increases further due to cell retention in the carriers. The highest value determined here is approx. 3–4 times higher than the maximum in chemostat cultivation. Obviously the maximum for fixed-bed cultures has not been reached so far.

All fixed-bed systems used here can be described by the same spline. This is very important with respect to scale-up, as obviously data from small-scale systems can be used to predict the performance on a larger scale (for more details on scale-up, see [20]).

4.3 Reaction kinetic model for start-up of fixed-bed reactors

For establishment of mathematical process model, biomass formation, lactose consumption, and lactate production during start-up of fixed-bed cultures with immobilized *L. lactis* were investigated experimentally and described by a reaction kinetic model [21]. Appropriate modeling and simulation of fixed-bed processes require biomass data. Therefore, a low-volume multiple fixed-bed reactor system (*Multiferm*) was used to investigate biomass formation of a *L. lactis* strain during the start-up phase of fixed-bed cultivation. The generation of data in parallel experiments was fast and easily compared to larger single reactor systems. Biomass data obtained from both fractions, retained and free suspended biomass, was used for modeling and simulation, together with data for lactose and lactate. The underlying Luedeking-Piret-like model structure was developed based on the results from suspension cultivations with the same strain. The fixed-bed system was described as perfusion culture with cell retention (**Figure** 7). For this, merely four additional parameters had to be defined to extend the suspension model to fixed-bed cultures. Experimental trends and steady states of both biomass fractions besides substrate and product could be described very well. Thus, this model could be used for process layout during process development.

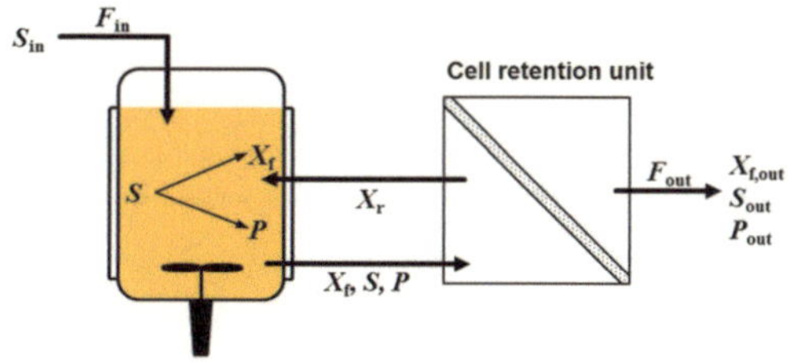

$$\mu = \mu_{max}\frac{S}{K_S + S}$$

$$\frac{dX_f}{dt} = X_f \cdot \mu - D \cdot (1 - R) \cdot X_f$$

$$\frac{dX_r}{dt} = D \cdot R \cdot X_f - k_{lys} \cdot X_r$$

$$\frac{dS}{dt} = -\frac{X_f \cdot \mu}{Y_{x/s}} - m_r \cdot X_r + D \cdot (S_{in} - S)$$

$$\frac{dP}{dt} = \alpha \cdot X_f \cdot \mu + \beta \cdot X_f - \gamma \cdot X_f \cdot S + \beta \cdot X_r \cdot \eta_P - D \cdot P$$

Parameter from suspension culture (chemostat)

μ_{max} (h^{-1})	K_s ($g \cdot L^{-1}$)	$Y_{x/s}$ ($g \cdot g^{-1}$)	α ($g \cdot g^{-1}$)	β (h^{-1})	γ ($L \cdot g^{-1} \cdot h^{-1}$)
0.48	0.24	0.18	10.22	0.12	0.53

Parameter from fixed bed culture (continuous)

R (-)	k_{lys} (h^{-1})	m_r (h^{-1})	η_p (-)
0.37	$2.92 \cdot 10^{-3}$	0.21	6.16

Figure 7.
Reaction kinetic model for fixed-bed cultures with immobilized microorganisms. For details see text and [21].

5. Conclusions

The goal of the studies was to evaluate the performance of fixed-bed bioreactor systems on different scales compared to suspension culture. The suggested concept for development of fixed-bed processes could be confirmed. The multi-fixed-bed bioreactor *Multiferm* provides an ideal downscaled and economical system that can be used for basic studies with low requirements on medium and cells. Here, questions such as optimal carrier design, appropriate medium, and process parameters (e.g., technique for immobilization, initial cell density, flow rate, temperature, oxygen, pH) can be evaluated. Especially the start-up phase can be investigated. The next step, a 100 mL fixed-bed system, provides data on the performance and long-term stability of the culture. Problems that might not have been shown up in the *Multiferm*, e.g., insufficient long-term stability, can be detected here. The 1 L fixed-bed can be regarded as a pilot scale already because medium requirement was already at 27.6 L per day at the highest dilution rate. Additionally, the radial-flow geometry can be easily scaled up further.

As expected, fixed-bed bioreactors could be operated in a perfusion mode at a steady state with dilution rates higher than the maximum specific growth rate. By this, very high volume-specific productivity with respect to lactate can be reached and maintained for long periods of time. The fixed-bed processes with lactic acid bacteria on macroporous carriers could be transferred on a pilot scale without loss in productivity. Furthermore, the productivity could be described by a spline, indicating that the maximum growth rate was not reached in this study.

Therefore, a process development tool for fixed-bed processes is now at hand that will pave the way for an industrial application of this promising technology.

Abbreviations and symbols

D	Dilution rate (h^{-1})
F	Flow rate ($L \cdot h^{-1}$)
F_{in}	Inlet flow rate ($L \cdot h^{-1}$)

F_{out}	Outlet flow rate ($L \cdot h^{-1}$)
j	Running index
k_{lys}	Lysis rate (h^{-1})
K_S	Substrate saturation constant ($g \cdot L^{-1}$)
LAB	Lactic acid bacteria
m_r	Maintenance rate of retained cells (h^{-1})
P_{out}	Product concentration at the outlet ($g \cdot L^{-1}$)
R	Fraction of retained biomass in perfusion fermentation (–)
S	Substrate concentration ($g \cdot L^{-1}$)
S_{in}	Substrate concentration at the inlet ($g \cdot L^{-1}$)
S_{out}	Substrate concentration at the outlet ($g \cdot L^{-1}$)
X_f	Free suspended biomass concentration ($g \cdot L^{-1}$)
$X_{f,out}$	Free suspended biomass concentration at the outlet ($g \cdot L^{-1}$)
X_r	Retained biomass concentration ($g \cdot L^{-1}$)
Y_j	Concentration of either biomass, substrate, or product ($g \cdot L^{-1}$)
$Y_{X/S}$	Biomass yield coefficient from substrate ($g \cdot g^{-1}$)
α	Growth-associated product formation rate ($g \cdot g^{-1}$)
β	Nongrowth-associated product formation rate (h^{-1})
γ	Constant for unspecific substrate loss ($L \cdot g^{-1} \cdot h^{-1}$)
μ	Specific growth rate (h^{-1})
μ_{max}	Maximum specific growth rate (h^{-1})
η_P	Effectiveness factor production (–)

Author details

Ralf Pörtner* and Rebecca Faschian
Institute of Bioprocess and Biosystems Engineering, Hamburg University of Technology, Hamburg, Germany

*Address all correspondence to: poertner@tuhh.de

References

[1] Tramper J. Conversions by immobilized cells versus traditional fermentations. In: Bont JAM, Visser J, Mattiasson B, Tramper J, editors. Physiology of Immobilized Cells. Amsterdam: Elsevier Science Publishers; 1989. pp. 115-126

[2] Eş I, Vieira JDG, Amaral AC. Principles, techniques, and applications of biocatalyst immobilization for industrial application. Applied Microbiology and Biotechnology. 2015; **99**(5):2065-2082. DOI: 10.1007/s00253-015-6390-y

[3] Polakovič M, Švitel J, Bučko M, Filip J, Neděla V, Ansorge-Schumacher MB, et al. Progress in biocatalysis with immobilized viable whole cells: Systems development, reaction engineering and applications. Biotechnology Letters. 2017;**39**(5):667-683. DOI: 10.1007/s10529-017-2300-y

[4] Sen P, Nath A, Bhattacharjee C. Packed-bed bioreactor and its application in dairy, food and beverage industry. In: Larroche C, Sanroman MA, Du G, Pandey A, editors. Current Developments in Biotechnology and Bioengineering: Bioprocesses, Bioreactors and Controls. Amsterdam: Elsevier; 2017. DOI: 10.1016/B978-0-444-63663-8.00009-4

[5] Pörtner R, Märkl H. Festbettreaktoren für die Kultivierung tierischer Zellen. BIOforum. 1995;**18**: 449-452

[6] Pörtner R, Faschian R, Goelling D. Fermentation of lactic acid bacteria: State of the art and new perspectives. In: Hilterhaus L, Liese A, Kettling U, Antranikian G, editors. Applied Biocatalysis. From Fundamental Science to Industrial Applications. Weinheim: Wiley-VCH Verlag GmbH; 2016, ISBN 978-3-527-33669-2

[7] Pörtner R, Seemuk M, Schlothauer R-C, Elsser D. Anaerobe Kultivierung von *Lactococcus lactis* im Festbettreaktor. Chemie Ingenieur Technik. 2004; **76**(10):1599-1602

[8] Fassnacht D, Pörtner R. Experimental and theoretical considerations on oxygen supply for animal cell growth in fixed bed reactors. Journal of Biotechnology. 1999;**72**(3):169-184

[9] Faschian R, De S, Pörtner R. Multi-fixed-bed bioreactor system applied for bioprocess development of immobilized lactic acid bacteria. The Open Biotechnology Journal. 2016;**10**:1-9. DOI: 10.2174/1874070701610010001

[10] Faschian R, Eren I, Minden S, Pörtner R. Evaluation of fixed-bed cultures with immobilized *Lactococcus lactis ssp. lactis* on different scales. The Open Biotechnology Journal. 2017;**11**: 16-25. DOI: 10.2174/18740707017 11010016

[11] Pörtner R, Platas OB, Fassnacht D, Nehring D, Czermak P, Märkl H. Fixed-bed reactors for the cultivation of mammalian cells: Design, performance and scale-up. Open Biotechnology Journal. 2007;**1**:41-46

[12] Atkinson B, Black GM, Pinches A. Process intensification using cell support systems. Process Biochemistry. 1980;**15**:24-32

[13] Mayo B, Aleksandrzak-Piekarczyk T, Fernndez M, Kowalczyk M, et al. Updates in the metabolism of lactic acid bacteria. In: Mozzi F, Raya RR, Vignolo GM, editors. Biotechnology of Lactic Acid Bacteria: Novel Applications. Oxford, U.K: Wiley-Blackwell; 2010. pp. 3-33

[14] Chahal SP. Lactic acid. In: Ullmann's Encyclopedia of Industrial Chemistry. Weinheim: Wiley-VCH; 2002

[15] Wee YJ, Kim JN, Ryu HW. Biotechnological production of lactic acid and its recent applications. Food Technology and Biotechnology. 2006; **44**(2):163-172

[16] Zhang Y, Cong W, Shi SY. Application of a pH feedback-controlled substrate feeding method in lactic acid production. Applied Biochemistry and Biotechnology. 2010;**162**(8):2149-2156

[17] Bai DM, Wei Q, Yan ZH, Zhao XM, Li XG, Xu SM. Fed-batch fermentation of *Lactobacillus lactis* for hyper-production of L-lactic acid. Biotechnology Letters. 2003;**25**(21): 1833-1835

[18] Ding S, Tan T. L-lactic acid production by *Lactobacillus casei* fermentation using different fed-batch feeding strategies. Process Biochemistry. 2006;**41**:1451-1454

[19] Lacroix C, Grattepanche F, Doleyres Y, Bergmaier D. Immobilised cell technologies for the dairy industry. In: Nedović V, Willaert R, editors. Applications of Cell Immobilisation Biotechnology. Heidelberg: Springer; 2005. pp. 295-319

[20] Hass V, Pörtner R. Praxis der Bioprozesstechnik. 2nd ed. Heidelberg, Germany: Spektrum Akademischer Verlag; 2011

[21] Faschian R, Minden S, Pörtner R. Experimental investigation and reaction kinetics modeling of biomass formation, substrate consumption and product formation during start-up of fixed-bed cultures with immobilized *Lactococcus lactis ssp. Lactis*. Journal of Bioprocessing and Biotechniques. 2016;**6**:9. DOI: 10.4172/2155-9821.1000294

Chapter 5

What Is Limulus Amebocyte Lysate (LAL) and Its Applicability in Endotoxin Quantification of Pharma Products

Yasir Mehmood

Abstract

Limulus amebocyte lysate (LAL) is an aqueous extract of blood cells (amoebocytes) from the horseshoe crab, *Limulus polyphemus*. LAL reagent reacts with bacterial endotoxin and lipopolysaccharide (LPS), which is a membrane constituent of Gram-negative bacteria. This reaction is the base on the LAL reagent, which is then used for the finding and quantification of bacterial endotoxins. The Gel Clot LAL test provides very simple positive or negative result and is most often mentioned in international pharmacopeia monographs as the official test. Gel Clot assay is a qualitative LAL test for detection of Gram-negative bacteria endotoxins. The Gel Clot assay is run in tubes that are placed in a water bath or in dry heated oven at 37°C. After a one-hour incubation period, the tubes are flipped 180°. A firm clot that stays in the bottom of the tube indicates a positive reaction. If the liquid flows down the side of the tube, the result is negative for endotoxins.

Keywords: Limulus amebocyte lysate, lipopolysaccharide, endotoxin, blood, bacteria, detection, horseshoe crab, pharmacopeias, delta, toxin, gel, chromogenic, acetic acid, Gram-negative

1. Introduction

Endotoxins, a type of pyrogen, are natural compounds found in the outer cell membrane of Gram-negative bacteria and can impact over 30 biological activities. Endotoxin can lead to cell death by initiating complement activation. The Limulus amebocyte lysate (LAL) test was commercially introduced in the 1970s. LAL is derived from the blood cells, or amebocytes, of the horseshoe crab, *Limulus polyphemus*. Frederick Bang and Jack Levin observed that blood cells from horseshoe crabs were found to clot in the presence of endotoxin, and this technology was used in the development of endotoxin detection assays. Today, endotoxin tests are performed on raw and in-process materials, and for the final release of products in the pharmaceutical and medical device industries.

Limulus amebocyte lysate test is an aqueous extract of blood cells (amoebocytes) which obtain from the horseshoe crab (*Limulus polyphemus*). LAL reagent reacts with the bacterial endotoxins or lipopolysaccharide (LPS). LAL test is recommended in all international pharmacopeias as the method for finding bacterial endotoxins. Gram-negative bacteria produce endotoxins (pyrogen). Exceptionally *Bacillus thuringiensis*, a Gram-positive bacteria produce delta toxin as endotoxins [1] (**Figures 1** and **2**).

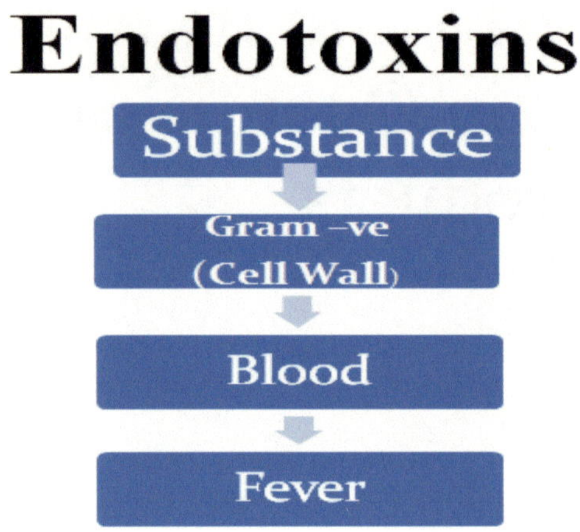

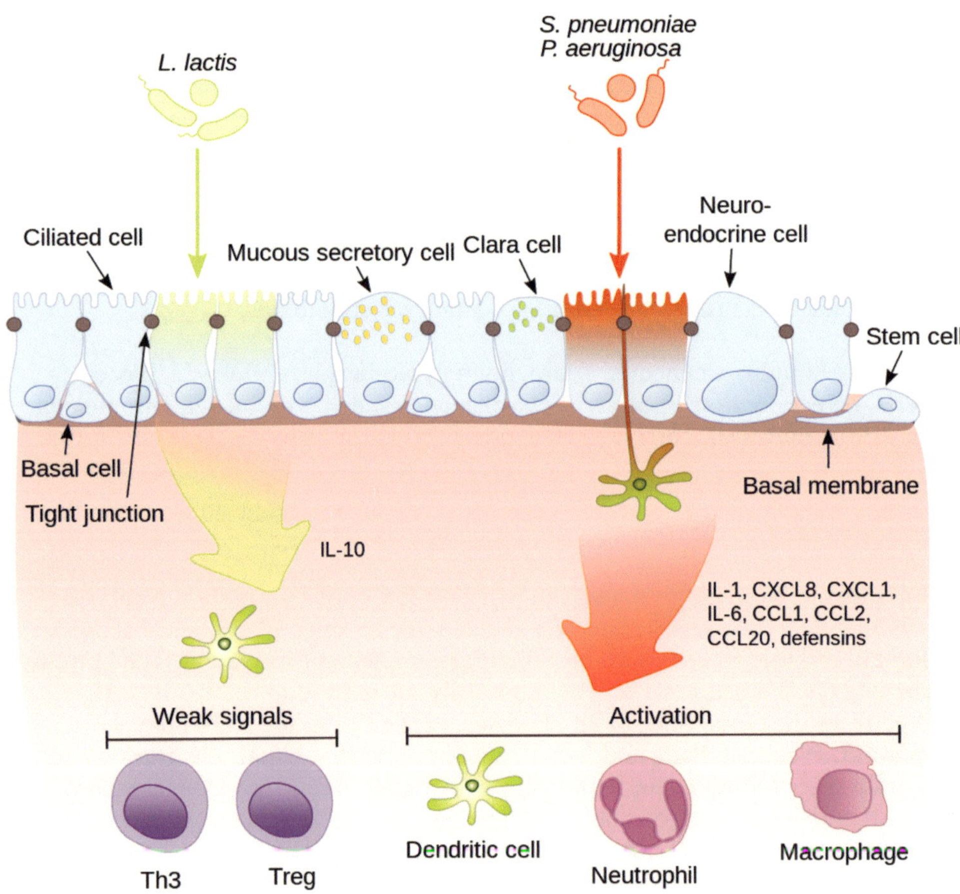

Figure 1.
Activation of inflammation in body [1]. Note: *LPS, lipoglycan. LAL test used according to the U.S. Food and Drug Administration (FDA) [2] guidelines Substituted for the U.S. Pharmacopeia (USP) pyrogen test (rabbit fever test) European Pharmacopeia (EP) Japanese Pharmacopeia (JP) [3]. LAL is used for human injectable drugs, animal injectable drug, medical devices, raw materials used in production, in process quality control.*

Figure 2.
Horseshoe crab [1].

2. Applications of the LAL test in the pharmaceutical industry

Among the most well-known and important applications of the LAL test are the ones related to the pharmaceutical industry. It can be said that the most common pyrogens in pharmaceutical products are endotoxins, which is why the pyrogen tests on rabbits have been replaced by the LAL test according to the recommendations of the international pharmacopeia. One of the reasons that has made the LAL test prevail in the pharmaceutical industry is the careful avoidance by the LAL manufacturers of bringing harm to live animals during both production and testing. It is important to clarify that the crabs, from which part of the hemolymph used for the LAL test was extracted, are returned to alive to their natural habitat with no lasting problems after the extraction.

2.1 Limulus amebocyte lysate (LAL) test types

Gel Clot technique: based on gel formation [4].

3. Methods to determine the pyrogen in pharma products

Chromogenic method: based on the producing color after cleavage of a synthetic peptide-chromogen complex.

Turbidimetric method: based on forming turbidity after cleavage of an endogenous substrate.

End point method: 0.005 endotoxins units (EU) per ml.

Kinetic method: 0.001 endotoxins units (EU) per ml.

Kinetic method: time taken to reach a specific absorbance at 405 nm (onset time) is determined. The assay requires specialized instrumentation. Take optical density readings at regular intervals. The greatest sensitivity, λ, of lysate is 0.001 EU/ml.

Endpoint chromogenic method [5]: the released amount of pNA can be calculated after a fixed incubation period. A standard curve, consisting of measured optical density plotted against known standard endotoxin concentration. Later used to determine concentrations in the product. The greatest sensitivity, λ, is 0.005 EU/ml UV/visible spectrophotometers.

4. Test performance

Add volume of lysate to a volume of product dilution. Incubating the reaction mixture at 37.5°C. Endotoxin in the reaction would activate the LAL reagent. Cleave small chromogenic peptides and liberates pNA. pNA, color is yellow and absorbs light at 405 nm. For samples that absorb at 405–410 nm, Diazo-coupling agent modification may be used. In this method, pNA reacted with nitrite in hydrochloric acid, ammonium sulfamate and N-(1-naphthyl)-ethylenediamine (NEDA). Absorbs at a range between 540 and 550 nm. A standard curve is used to establish concentrations in product specimens.

5. Materials and equipment

10 × 75 mm fully depyrogenated borosilicate glass culture tubes (Associates of Cape Cod, Inc. catalog numbers TB050).

Optical reader is capable of reading at 405 nm, or at 540–550 nm for the diazo method. Incubator is able to maintaining 37 ± 1°C. A water bath can be used for the endpoint test tube method. Both devices should have a uniform heat distribution. Test tube racks to hold the tubes and/or incubate dilution and reaction tubes. Micropipettes or disposable pipette tips free of interfering endotoxins and glucans are recommended. Vortex-type mixer, Para film (American National Can™) and hot-air oven with the capacity to heat to at least 250°C for depyrogenation of glassware.

6. Chemicals and reagents

Limulus amebocyte lysate (LAL), LAL reconstitution buffer, control standard endotoxins (CSE), solution 1 (nitrite), solution 1A (0.1 N hydrochloric acid), solution 2 (ammonium sulfamate), solution 3 (N-(1-naphthyl)-ethylenediamine (NEDA)), LRW.

The endotoxins limit for USP/BP sterile WFI is only 0.25 EU/ml; therefore, sterile WFI may contain detectable endotoxins and be unsuitable for use. Use certified LRW to make dilutions of standards, and to prepare positive controls.

7. Quality control steps or test procedure

7.1 Specimen collection and preparation

Collect aseptically containers that are free of detectable endotoxins in depyrogenated glassware apparatus.

7.2 pH of the specimen

The pH must be 6–8. Adjust the pH of the product specimen with dilute HCl, NaOH, or buffer (free of endotoxins). Dilute concentrated HCl or NaOH with LRW. Use a volume that will not lead to significant dilution of the test specimen. Dilution (LRW) alone can overcome the issue sometimes.

7.3 Method of lysate reconstitution

Gently tap the vial of lysate. Loose material fall to the bottom. Break the vacuum by lifting the gray stopper. Do not contaminate the mouth of the vial. Remove and discard the stopper. Start the reconstituted lysate with 3.2 ml buffer. Avoid vigorous mixing that may cause excessive foaming and a loss of sensitivity. Wrap the vials with parafilm and store in a cold place (2–8°C) when not in use and use within 8 h of reconstitution.

7.4 Lysate storage conditions

7.4.1 Lyophilized lysate

This is relatively well stable and, if stored properly, will retain full activity through the expiration date on the label. Store the product at 2–8°C. Excess temperature over 37°C cause rapid deterioration, loss of sensitivity and distinct yellowing.

7.5 Control standard endotoxins (CSE)

Each vial of control standard endotoxins (CSE) contains 10 ng of endotoxins. Reconstitute CSE with the volume mentioned on the Certificate of Analysis (CA, which gives the potency of the CSE). Gently knocks the vial of control standard endotoxins (CSE) to cause loose material to fall to the bottom. Break the vacuum by lifting the gray stopper. Do not contaminate the mouth of the vial. Remove the stopper and place it in a cold place aseptically for reuse.

Reconstitute CSE with the volume specified on the Certificate of Analysis (CA, which gives the potency of the CSE) and as directed in the package insert. Place the stopper. Vortex the vial for 40–60 s to form a homogenous mixture. Discard solution if not used immediately, vortex the vial for 30 s prior to use.

7.5.1 Mixing and incubation

Read the tubes UV/visible spectrophotometers (**Table 1**).

CSE + lysate	Incubation time (min)
50 μl of 0.50 EU/ml + 50 μl	30
50 μl of 0.250 EU/ml + 50 μl	30
50 μl of 0.125 EU/ml + 50 μl	30
50 μl of 0.0625 EU/ml + 50 μl	30

Table 1.
Dilution mixing and incubation time.

7.5.2 Mixing and incubation

Stop the reaction by adding 50% acetic acid. Add 0.025 ml (25 μl) read the optical density (OD) at 405 nm read the test.

Sample + lysate	Incubation (min)
50 μl of sample + 50 μl	30

7.6 Stop reaction solution preparation

7.6.1 Read the test

Reconstitute vial 1 with entire contents of vial, reconstitute vial 2 with 4 ml of water, reconstitute vial 3 with 4 ml of water. Add 0.05 ml (50 μl) of solution 1 (sodium nitrite reconstituted with dilute HCL). Add 0.05 ml (50 µl) of solution 2 (ammonium sulfamate). Add 0.05 ml (50 µl) of solution 3 (NEDA) use new pipette tip agitate the plate to mix. Full color (magenta) should develop immediately. Read the test at 540–550 nm.

7.6.2 Positive control

7.6.2.1 Make standard curve

Positive control must be included to verify that it is appropriate to use the parameters of a previous (archived) standard curve to calculate endotoxin concentrations.

7.6.3 Negative controls

LRW negative controls should be included with each test

1: Equation of straight line (results)

$y = mx + c$

m = slop

x = endotoxin concentration,

c = y-intercept and

y = mean absorbance

$X = y\text{-}c/m$

Example calculation

7.6.4 Sample preparation

Prepare sample solutions by dissolving or diluting drugs (pH 6.0–8.0). The pH may be adjusted by the use of acid, base, or suitable buffers as recommended. Do not exceed the MVD or MCV while making dilutions and adjusting the pH.

7.6.5 Maximum valid dilution (MVD)

MVD = (endotoxin limit × concentration of sample solution)/(λ)

Endotoxin limit given in USP, concentration of a sample of the label, λ: the labeled lysate sensitivity in the gel-clot technique (IU/ml) or the lowest concentration used in the standard curve for the turbidimetric or chromogenic techniques.

7.6.6 Minimum valid concentration (MVC)

MVC = λ/endotoxin limit

λ: the labeled lysate sensitivity in the gel-clot technique (IU/ml) or the lowest concentration used in the standard curve for the turbidimetric or chromogenic techniques.

Sample 1

Endotoxin limit: 0.5 EU/ml

Concentration of sample: 100 mg/ml

λ: 0.06 EU/ml

MVD = 0.5 EU/ml × 100 mg/ml/0.06 EU/ml

MVD = 833

Add 1 ml of sample 1 in to 832 ml of LRW. Prepare sample 2 in using the same method.

7.6.7 Preparation of CSE dilutions

Using 10-fold and 2-fold dilution methods prepare the following dilutions of control standard endotoxins (CSE)

1. 0.5 EU/ml

2. 0.25 EU/ml

3. 0.125 EU/ml

4. 0.0625 EU/ml

Reconstitute the lysate with 3.2 ml of buffer provided with it. Follow the standard procedure for reconstitution.

7.6.7.1 Mixing and incubation

Stop reaction.
For sample 1 and sample 2:
Stop the reaction by adding 50% acetic acid. Add 0.025 ml (25 μl) (**Tables 2** and **3**).

CSE + lysate	Incubation (min)
50 μl of 0.50 EU/ml + 50 μl	30
50 μl of 0.250 EU/ml + 50 μl	30
50 μl of 0.125 EU/ml + 50 μl	30
50 μl of 0.0625 EU/ml + 50 μl	30
50 μl of sample 1 + 50 μl	30
50 μl of sample 2 + 50 μl	30

Table 2.
Different dilution of CSE and lysate.

Sample	Absorbance at 405 nm	Mean Absorbance	Mean Absorbance − Blank Absorbance
LAL Reagent Water (Blank)	0.080	0.082	
	0.084		
0.0625 EU/ml	0.128	0.126	0.044
	0.124		
0.125 EU/ml	0.160	0.170	0.088
	0.180		
0.25 EU/ml	0.309	0.317	0.235
	0.325		
0.5 EU/ml	0.570	0.564	0.482
	0.557		
Sample 1	0.372	0.382	0.300
	0.392		
Sample 2	0.416	0.417	0.335
	0.418		

Table 3.
Make two replicates of each CSE and sample preparation to reduce any errors.

8. Results

Use Microsoft word for further calculations and results. Make standard curve and endotoxin concentration (**Figure 3**).

R^2 = coefficient of determination
R = correlation coefficient
$R \geq 0.98$
$R^2 = 0.99$
$R = \sqrt{R^2} = 0.99$
Equation of straight line
y = mx + c
m = slop
x = endotoxin concentration
c = y intercept
y = mean absorbance
Equation of straight line
Y = 1.019X – 0.026

8.1 Rearranging the equation

X = Y + 0.026/1.019
m = slop = 1.019,
C = y intercept = 0.026,
Y = mean absorbance,
X = endotoxin concentration

8.2 Sample 1

X = Y + 0.026/1.019
Y = 0.300, X = 0.300 + 0.026/1.019, X = 0.319EU/ml

8.3 Sample 2

X = Y + 0.026/1.019
Y = 0.335, X = 0.335 + 0.026/1.019, X = 0.354 EU/ml

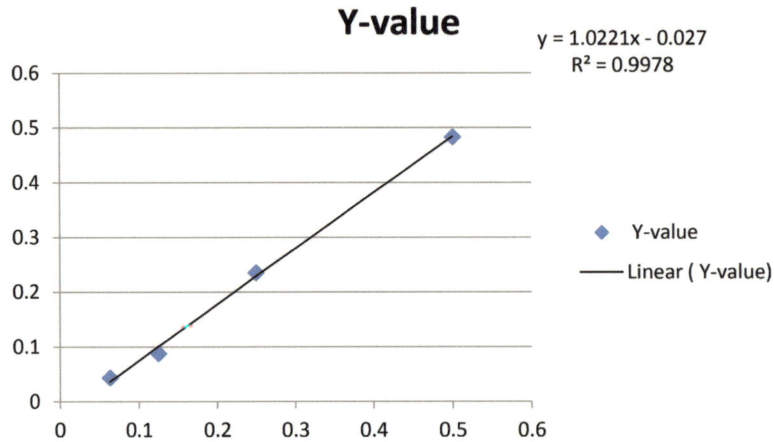

Figure 3.
Validation of standard curve.

8.4 Advantages of Gel Clot method

Gel Clot LAL provides a simple positive/negative result and is most often mentioned in pharmacopeial monographs as the official referee test.

This is very easy to perform.

This is not time consuming.

Accuracy is 100 percent.

The LAL Gel-Clot assay, gives a more quantitative measurement of endotoxin over a range of concentrations.

8.5 Standard operating procedure

8.5.1 Material

Gel Clot lysate for 20 test, Gel Clot standard 0.5 EU/Vial, LAL reagent water (LRW 50 ml).

8.5.2 Reconstitution

Lysate: add 2 mL LRW and mix it slowly. Do not shake and avoid foaming. Transfer 0.1 ml in 20 test tubes. Store it at –degree (in freezer).

Standard: Add 2 mL of LRW in the vial and mix it well for 15 min. Store the vial at 2–8°C. Storage life is 15 days.

8.5.3 Procedure

Take three test tubes and mark them as test, positive control and negative control [1].

Add your sample in test tube marked as sample. Add standard in test tube marked as Positive control. Add LRW in test tube marked as negative control. Incubate the test tubes at 37 + 2°C for 60 min. After an incubation, check for the gel by inverting the test tube. If the material remains firm in the bottom of the test tube, it means gel has formed. This positive if the material gets the flow down, it means gel has not formed. This means negative.

8.5.4 For water for injection

Take similarly three test tubes as above and add water for injection (WFI) in test tube marked as sample. And proceed as above. The results should be as follows (**Table 4**):

8.5.5 For product

We have to make dilution.

Sample	Positive control	Negative control	Result
−ve (gel not formed)	+ve	−ve	Pass
Sample	**Positive control**	**Negative control**	**Result**
+ve (gel formed)	+ve	−ve	Fail

Table 4.
Results shown sample pass or not.

Example: If the product endotoxin limit is 1 EU/ml, then we have to make the dilution as follows:

Since we are using 0.25 EU/ml, this is called lambda. Divide the endotoxin limit of product with lambda

1/0.25 = 1:4

As per USP, we have to test 3 test as follows:

One test tube	1:3	The result should be positive
Second	1:4	The result may be positive or negative
Third	1:5	The result should be negative

This means the product is passed.

8.5.6 LAL test reagents (chromogenic method)

Chromogenic lysate [2],
Respective endotoxin standard,
Diazo coupling reagent (set of four bottles).
Note: All reagents must be stored in refrigerator at 2–8°C.

8.5.7 Preparation of acetic acid 0.8 m

Dissolve 45.6 ml of acetic acid in 1 liter of distilled water. The final concentration of acetic acid is 0.8 M. This solution can be stored for 3 months.

Remove the plastic cover. Wipe off with 70% alcohol around the rubber cap and top portion of every vial. Remove the aluminum cap with sterile and pyrogen free forceps and then cover with depyrogenated aluminum foil to avoid any Endotoxin contamination. (2.8 ml LAL water vial is provided with Endotoxin vial, concentration is mentioned on the label). Pour whole quantity of LAL water into the ET vial and cover with foil. Mix vigorously for at least 10 s by vortexer. During stirring solution must not touch the foil.

Storage: Store reconstituted Endotoxins solution at 4°C in a refrigerator for 14 days. The solution can also be stored at –20°C for a month. Avoid freezing during storage.

Note: Stir every time vigorously before use.

Toxicolor lysate

(Buffer vial 0.35 ml and LAL water are provided with Lysate. Sensitivity is mentioned on the certificate). After taking from the refrigerator, pour whole quantity of buffer and 0.35 ml LAL water into the lysate vial as soon as possible, covers with foil. Then quickly stir to dissolve. Avoid air bubbling during stirring. Place the vial in ice water bath for 2–3 min before use.

Note: Be sure that the reagent is completely dissolved. This reagent must be reconstituted just before use. The reagent is extremely sensitive and must be consumed at one time. Storage should be avoided, but can be stored at −20°C in 0.1 ml dispensed quantities in small test tubes. Use stored lysate if the color is not changed. Reconstituted lysate may only be deep frozen once.

8.5.8 Diazo coupling reagent

Four bottles are provided with one set, marked as 7, 8, 9 and 10s respectively. Transfer whole quantity of bottle no. 7 s into bottle no. 8 s. Then add 12 ml distilled water into each of bottle no. 9 s and 10s. Ultimately, we will have three bottles 8, 9, and 10 s, which are used stepwise to block the reaction.

9. Pre-test preparations

9.1 pH of the sample

The pH of the sample is adjusted by pyrogen free 0.1 N NaOH or 0.1 N HCl. The pH of the sample should be between 6.0 and 8.0.

9.2 Test tubes settings

Arrange test tubes in two stands as under; stand 1—test tubes for sample and standard dilutions; stand 2—test tubes for reaction.

10. Procedure

10.1 Sample preparations

Take 0.05 ml well-mixed sample into small test tubes. If required, make 1/10 dilution of the sample with Pyrogen free water as Below, Take 4.5 ml of pyrogen free water in the test tube. Then add 0.5 ml of well-mixed sample. Vortex mixing for a few seconds.

Take 0.1 ml into a small test tube for further process.

10.2 Standard preparation

Make a dilution of the endotoxin (concentration 0.470 EU/ml) according to the product limit. For making 0.235 EU/ml (if the product limit is 0.25) proceed as follows;

Take 0.05 ml of the reconstituted endotoxin in the test tube after stirring. Add 0.05 ml of pyrogen free water and vortex to mix. Now the final dilution is 0.235 EU/ml.

Take 0.05 ml of step 2 into a small test tube for further process.

10.3 Blank preparation

Pour 0.05 ml of pyrogen free water (being used in the test) in small test tube as a blank for further process.

Lysate addition

Place the tube stand for small test tubes (containing the tubes of blank, standard and diluted samples) in ice water bath or suitable ice water container. Add 0.05 ml of lysate to all of the tubes as soon as possible. Stir the contents of every tube soon after the addition of lysate for a few seconds. Avoid foaming.

10.4 Incubation

Soon after the addition of lysate, place the test tube rack in the incubator set at 32.5 + 2.5°C for 30 min. The tube rack can be placed in the water container placed in the incubator.

10.5 Blocking the reaction

After completion of the incubation period, place tube rack in ice water bath, then blocks the reaction immediately from one of the two methods mentioned below:

By acetic acid

Add 0.4 ml of 0.8 M acetic acid into each tube and stir to mix.

By diazo coupling reagent

Three bottles of the reagent are used as under;

Add 0.5 ml from bottle no 8 s to each tube and stir to mix.

Add 0.5 ml from bottle no 9 s to each tube and stir to mix.

Add 0.5 ml from bottle no 10s to each tube and stir to mix.

Absorbance reading (using spectrophotometer) measurement at 405 nm

If 0.8 M acetic acid is used to block the reaction, then absorbance reading is taken at 405 nm.

Note: The readings. Glass photocell is used for reading at 405 nm. Because the volume of the tube content is not sufficient, the distilled water is added to each tube and is stirred to mix.

10.6 Measurement at 545 nm

When Diazo coupling reagent is used for blockage of the reaction then the reading is taken at 545 nm. Note all the readings.

Note: Distilled water is used for reference in both cases.

11. Results from software

All the absorbance readings are fed in the "Software reader for window version 1.51" to collect the results.

11.1 Manual measurement of endotoxin

Following Formula is used to calculate the results

$$\text{Endotoxin/ml in product} = \frac{\text{conc.of std.in solution} \times \text{sample ABS} - \text{blank ABS}}{\text{Std.ABS} - \text{blank ABS}} \times \text{DF}$$

Calculations of MVC, MVD and ELC.

$$\text{MVC} = \text{M/K}$$

Where the lowest sensitivity of lysate, M is the maximum dose/kg body weight and K is constant having value equal to 5.

$$\text{MVD} = \text{concentration of product in 1 ml/MVC}$$

$$\text{ELC} = \text{X MVD}$$

Conflict of interest

The author(s) confirm that this chapter content has no conflict of interest.

Author details

Yasir Mehmood
Faculty of Pharmaceutical Sciences, Government College University Faisalabad, Pakistan

*Address all correspondence to: yasir_dpharm@hotmail.com

References

[1] Devleeschouwer M, Cornil M, Dony J. Studies on the sensitivity and specificity of the Limulus amebocyte lysate test and rabbit pyrogen assays. Applied and Environmental Microbiology. 1985;**50**(6):1509-1511

[2] Food, Administration D. Guideline on Validation of the Limulus Amebocyte Lysate Test as an End-Product Endotoxin Test for Human and Animal Parenteral Drugs, Biological Products, and Medical Devices. Center for Drug Evaluation and Research. Rockville, MD, USA: Food and Drug Administration; 1987

[3] Committee JPE. The Japanese Pharmacopoeia. Tokyo, Japan: The Society of Japanese Pharmacopoeia; 2006. p. 15

[4] Ong KG, Leland JM, Zeng K, Barrett G, Zourob M, Grimes CA. A rapid highly-sensitive endotoxin detection system. Biosensors and Bioelectronics. 2006;**21**(12):2270-2274

[5] Scully M, Newman Y, Clark S, Kakkar V. Evaluation of a chromogenic method for endotoxin measurement. Thrombosis Research. 1980;**20**(2):263-270

Chapter 6

The War between Bacteria and Bacteriophages

Beata Orzechowska and Manal Mohammed

Abstract

The rapid emergence and dissemination of multidrug-resistant (MDR) bacteria represents a worldwide crisis concerning that humankind is re-entering the 'pre-antibiotics' era. Before the discovery of antibiotics, bacteriophage therapy was widely enforced to combat bacterial infections. However, the discovery of penicillin in 1940 and other novel antibiotics replaced phage therapy, and they are being used as the first line of defence against pathogenic bacterial infections. Factors such as selective pressure resulted in bacteria becoming insensitive to one or multiple antibiotics, frequently leading to limited treatment options. This prompted a renewal of interest to the phage therapy that remains dubious due to its disadvantages such as host specificity and the development of bacterial resistance against phages. Evolution of bacterial genomes allowed bacteria to acquire vast mechanisms interfering with phage infection such as inhibition of phage adsorption, prevention of phage entry, superinfection exclusion, restriction-modification and abortive infection. Interestingly, phages have developed diverse counterstrategies to circumvent bacterial anti-phage mechanisms including digging for receptors, adapting to new receptors and masking and modifying restriction sites. Understanding the complex dynamics of bacteria-phage interaction is a preliminary step towards designing synthetic phages that can overcome limitations of phage therapy and potentially lead to defeating MDR bacteria.

Keywords: bacteria-phage arms race, CRISPR system, anti-CRISPR system, superinfection exclusion (Sie), restriction-modification, abortive infection (Abi)

1. Introduction

Antimicrobial resistance is a global public health crisis. According to Public Health England [1], each year approximately 25,000 people die across Europe due to hospital-acquired infections caused by antibiotic-resistant and MDR bacteria such as *Mycobacterium tuberculosis*, Methicillin-resistant *Staphylococcus aureus* and multiresistant Gram-negative bacteria. Gram-negative infections include those caused by *Escherichia coli*, *Klebsiella pneumoniae* and *Pseudomonas aeruginosa* [2]. Nevertheless, it is estimated that by 2050, the global yearly death toll will increase to 10 million. Accelerating emerge of antimicrobial resistance seriously threatens the effectiveness of treatments for pneumonia, meningitis and tuberculosis, in addition to diminishing prevention of infections acquired during surgeries and chemotherapies. The crisis of the antibiotic resistance requires urgent, coordinated action.

Misuse and overuse of antibiotics must be controlled, implementation of new policies regarding prescriptions has to be internationally addressed; and development of new therapeutics is urgently required [1].

Félix d'Herelle, known as the father of bacteriophage (or phage) therapy [3], brought an evolutionary discovery of phages as therapeutics for various infections and conditions. Phage therapy was widely enforced in the 1920s and 1930s to combat the bacterial infections. However, in the 1940s, the newly discovered antibiotics replaced the phage therapy (except Russia, Georgia and Poland) [4].

The emergence of MDR bacteria prompted a renewal of the interest to the phage therapy as an alternative treatment to overcome a broad spectrum of resistant bacterial infections. Phage therapy and phage cocktails that contain a mixture of different bacteria-specific phages, drawn interest within molecular biology and modern medical research as potential antimicrobials that could tackle the crisis of antimicrobial resistance. Nonetheless, the phage therapy remains controversial due to its disadvantages such as bacteriophage resistance: bacteria-phage evolutionary arms race that could put a burden on a long-time application of phage therapy as an anti-infectious agent [5].

Phage therapy has many advantages, primary because phages are very specific (generally limited to one species) and easy to obtain as they are widely distributed in locations populated by bacterial hosts including soil and seawater, and they do not have any known chemical side effects like antimicrobials [6].

Understanding host-phage interactions and 'the war between bacteria and phages' are steps towards designing engineering 'broad-spectrum phage' that can overcome the limitations of phage therapy and potentially overcome a wide range of resistant bacterial infections [6].

2. The evolutionary phage-host arms race

Phages are obligate intracellular parasites that distinctively infect bacterial cells. Although phages are very specific to their host, generally limited to one species, they pose an enormous threat to bacteria as in some habitats they outnumber their hosts by nearly 10-fold number [7]. Phages are the most abundant, ubiquitous and diversified organisms in the biosphere [8, 9]. Phage-host interaction and fight for the survival led to the evolution of bacterial and viral genomes and, therefore, to the evolution of resistance mechanisms. Bacteria, continuously, evolve many molecular mechanisms, driven by gene expression to prevent phage infection. These evolving phage-resistance mechanisms in bacteria induce the parallel co-evolution of phage diversity and adaptability [10, 11]. The co-evolving genetic variations and counteradaptations, in bacteria and phages, drive the evolutionary phage-host arm race [11, 12].

Leigh Van Valen, an evolutionary biologist, metaphorised the co-evolutionary arm race and proposed the Red Queen hypothesis [13].

'It takes all the running you can do, to stay in the same place' the Red Queen says to Alice in *Through the Looking-Glass*.

The Red Queen hypothesis proposes that to survive, microorganisms must constantly adapt, evolve and thrive against ever-evolving antagonistic microorganisms within the same ecological niche [14].

Bacteria have developed various anti-phage mechanisms including non-adaptive defences (non-specific) and adaptive defences associated with Clustered Regularly Interspaced Short Palindromic Repeats (CRISPR) along with CRISPR-associated (Cas) proteins [7, 15–18].

The non-specific adaptations (analogues to innate immunity in multicellular organisms) act as primary mechanisms to evade viral infection, and they include mechanisms that inhibit phage adsorption and prevent nucleic acid entry, superinfection exclusion systems, restriction-modification systems and abortive infection [7, 19].

On the other hand, the adaptive resistance (analogues to the acquired immunity in multicellular organisms) serves as a second line of defence, which is very efficient and phage-specific.

Interestingly, it was observed that the bacterial anti-phage mechanisms are generally present in a genomic array, known as 'defence islands' [20]. The 'defence islands' are enriched in putative operons and contain numerous overrepresented genes encoding diverged variants of antiviral defence systems. Moreover, scientific evidence and characteristic operonic organisation of 'defence islands' show that many more anti-phage mechanisms are yet to be discovered [21–24].

Although bacteria have developed several resistance mechanisms against phages, phages can circumvent bacterial anti-phage mechanisms on the grounds of their genomic plasticity and rapid replication rates. These counterstrategies include point mutations in specific genes and genome rearrangements that allow phages to evade bacterial antiviral systems such as CRISPR/Cas arrays by using anti-CRISPR proteins and abortive infection by hijacking bacterial antitoxins, as well as escaping from adsorption inhibition and restriction-modification mechanisms [15–18].

This chapter will comment on the genetic basis of bacterial resistance to phages and different strategies used by phages to evade bacterial resistance mechanisms.

3. Preventing phage adsorption and phage's counterstrategy

Phage adsorption to host-specific receptors on the cell surface is the initial step of the infection and host-phage interaction. Depending on the nature of bacteria, whether it is Gram-positive or Gram-negative proteins, lipopolysaccharides, teichoic acids and other cell surface structures can serve as irreversible phage-binding receptors [19]. These receptors might be present in the cell wall, bacterial capsules, slime layers, pili or flagella [25].

Bacteria have acquired various barriers to inhibit phage adsorption, such as blocking of phage receptors, production of extracellular matrix (e.g. capsule, slime layers) and production of competitive inhibitors [26–31]. The diversity of phage receptors in the host is influenced by co-evolutionary adaptations of phages to overcome these barriers [32]. This includes diversity-generating retroelements (DGRs) and phase variation mechanisms causing phenotypical differences within the bacterial colony [7, 33, 34].

Phase variation is a heritable, yet reversible process regulating gene expression in bacteria; genes can switch between a functional (expression) and a non-functional state leading to phenotypical variations within the bacterial population even when strains have identical genotype. Sørensen et al. [35] investigated the underlying resistance mechanism of *Campylobacter jejuni* (NCTC11168) to phage F336. They have discovered that phage F336 relies on the hypervariable O-methyl phosphoramidate (MeOPN) modification of capsular polysaccharides (CPS) for successful adsorption to the bacterial surface. Nevertheless, loss of MeOPN receptor on the bacterial cell surface due to phase variation in the *cj1421* gene encoding the MeOPN-GalfNAc transferase (MeOPN transferase attaches MeOPN to GalfNAc and Hep side chains of CPS) results in phage resistance [35, 36].

DGRs are genetic elements diversifying DNA sequences and the proteins they encode ultimately mediating the evolution of ligand-receptor interactions. Error-prone DGRs and random mutations in the bacterial genes encoding cell surface receptors lead to the alternation and change in the structural composition of the phage receptors, making them non-complementary to the phage's anti-receptors, known as receptor-binding proteins (RBP) [34] (**Figure 1(1)**).

Yet, phage's replication is exceedingly error-prone, therefore causing many random mutations in the genes encoding the RBP or tail fibres. Phages also possess DGRs that mediate phage's tropism by accelerating the variability in the receptor-coding genes through reverse transcription process [37]. The changes in the nucleotide sequence in the RBP-coding gene may ultimately lead to the adaptation to the modified receptor (**Figure 1(2)**), thus the ability to adsorb and infect the bacterial cell.

Unsurprisingly, bacteria also exhibit different strategies to block their receptors [28–31].

Figure 1(4) demonstrates the findings from studies conducted on *Staphylococcus aureus* by Nordstrom and Forsgren [38]. Mutants of *Staphylococcus aureus* producing higher anticomplementary protein A were found to adsorb fewer phages than *Staphylococcus aureus mutants with scarce of protein A, which had an apparent increased ability to adsorb phages* [38]. These findings indicate that some bacteria, including *Staphylococcus aureus, are capable of production of surface proteins* that mask the phage receptors making them inaccessible for phage recognition and attachment (**Figure 1(3)**).

Receptors located on bacterial cell surface serve a vital role in bacterial metabolism; they may function as membrane porins, adhesions or chemical receptors [19]. Therefore, mutation or complete loss of the receptor might be lethal for bacteria. To inhibit phage adsorption, bacteria can produce surface molecules, such as exopolysaccharides.

Exopolysaccharides are extracellular polysaccharides acting as a physical barrier, composing slime or capsules surrounding bacterial cells that lead to inaccessible host receptors for efficient phage adsorption [39] (**Figure 2**). Studies conducted by Looijesteijn et al. [40] shown that exopolysaccharides *produced by Lactococcus lactis function as external protection from phages and the* cell wall destructing lysozyme, due to masked cell surface receptors [40].

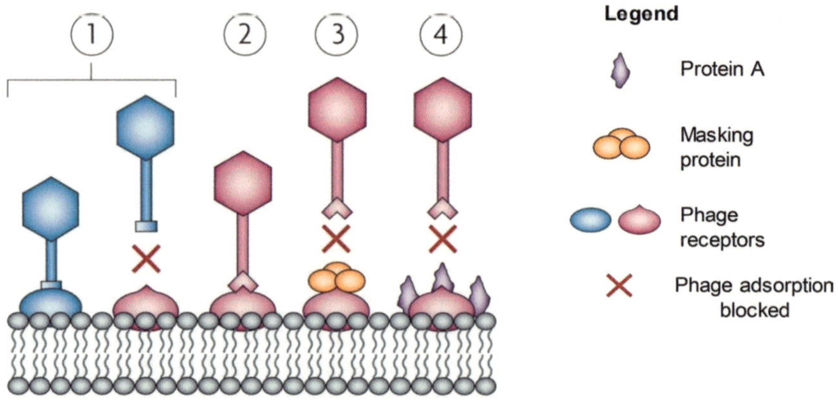

Figure 1.
Bacterial defence mechanisms preventing phage adsorption and phage's counteradaptations. (1) Phage adsorption to a host-specific receptor site on a host cell surface. Bacterium evolves phage resistance by the modification of these cell surface receptors; phage is incapable of binding to the altered receptor. (2) Phage's adaptation to these modifications through mutations in receptor-binding protein gene that leads to the co-evolution of bacterial genetic variation. Bacteria are also capable of producing proteins that mask the phage recognition site receptors (3 and 4), thus making the receptor inaccessible for phage adsorption [28–31]. Image courtesy of springer nature: https://www.ncbi.nlm.nih.gov/pubmed/20348932.

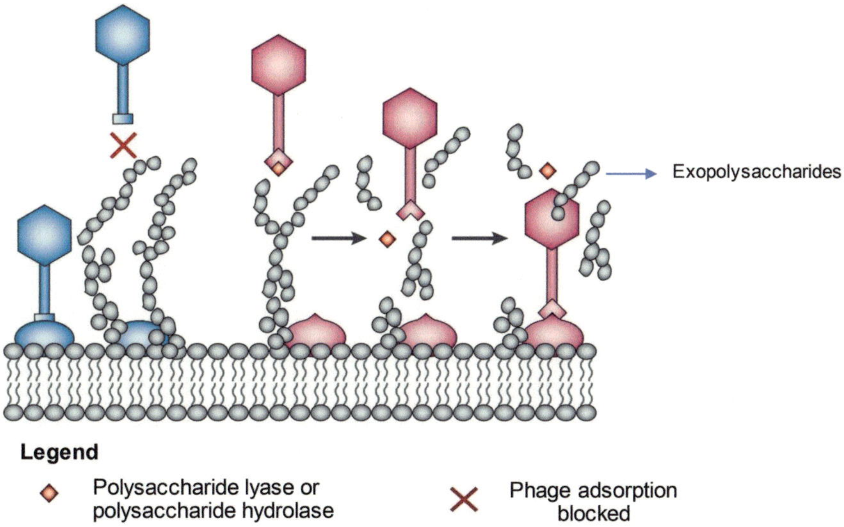

Figure 2.
Bacterial strategies to inhibit phage adsorption and phage strategies to access host receptors. Some bacteria are capable of the production of exopolysaccharides, which act as an outer shield, protecting a cell from the phage infection [28–31]. If the phage does not possess any polysaccharide-degrading enzymes, it cannot access the host cell membrane receptor. However, some phages evolved mechanisms allowing them to recognise these extracellular matrixes and degrade them by the means of hydrolases and lyases [15–18]. Image courtesy of Springer Nature: https://www.ncbi.nlm.nih.gov/pubmed/20348932.

Nevertheless, some phages evolved mechanisms allowing them to recognise these extracellular matrixes and degrade them by utilising hydrolases and lyases (**Figure 2**) [15–18]. The polysaccharide-degrading enzymes allow phages to gain access to the receptor that may lead to the viral propagation. They are commonly present bound to the RBPs or exist as free soluble enzymes from previously lysed bacterial cells [41].

4. Preventing phage DNA entry and phage's counteradaptations

If phage bypasses primary antiviral strategies, it is now able to initiate infection by adsorption to a specific receptor site on a host cell surface through phage RBP [42, 43]. Upon interaction with the cell receptors, the phage injects its genetic material (single or double-stranded DNA or RNA) into the cytoplasm of the host. Depending on the nature of the phage and growth conditions of the host cell, it follows one of the two life cycles: lytic or lysogenic (**Figure 3**).

In the lytic cycle, virulent phages degrade host's genome leading to the biosynthesis of viral proteins and nucleic acids for the assembly of phage progeny. Eventually, the bacterial cell lysis, releasing a multitude of newly assembled phages, is ready to infect a new host cell [46].

In contrast, temperate phages might enter the lytic or lysogenic cycle, if the host cell exists in adverse environmental conditions that could potentially limit the number of produced progeny (**Figure 3** demonstrates typical lifecycle of temperate phage using coliphage λ as an example) [44, 45]. In the lysogenic phase, repressed phage genome integrates into the bacterial chromosome as a prophage. This process causes the proliferation of prophage during replication and binary fission of bacterial DNA.

Prophage only expresses a repressor protein-coding gene. The repressor protein binds to the operator sites of the other genes and ultimately inhibits synthesis of phage enzymes and proteins required for the lytic cycle.

When the synthesis of the repressor protein stops or if it becomes inactivated, a prophage may excise from the bacterial chromosome, initiating a lytic cycle (induction) which leads to the multiplication and release of virulent phages and lysis of a host cell [44, 45].

If the phage remains in the nearly dormant state (prophage), the lysogenic bacterium is immune to subsequent infection by other phages that are the same or closely analogous to the integrated prophage by means of Superinfection exclusion (Sie) systems [47].

Sie systems are membrane-associated proteins, generally, phage or prophage encoded, that prevent phage genome entry into a host cell [47]. **Figure 4** shows the role of Sie system (proteins Imm and Sp) in blocking phage T4 DNA entry into Gram-negative *Escherichia coli*. Despite successful attachment to the phage-specific receptor, phage DNA is directly blocked by Imm protein from translocating into the cytoplasm of the cell. Sp system, on the other hand, prevents the degradation of the peptidoglycan layer by inhibiting the activity of T4 lysozyme [26–31, 48].

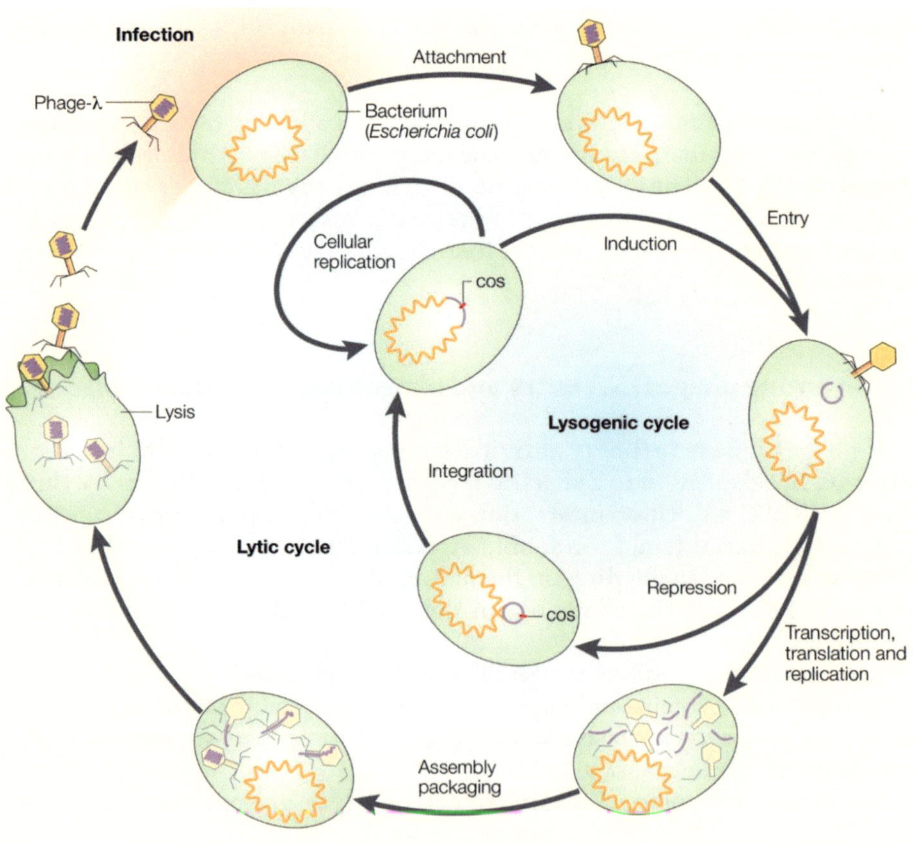

Figure 3.
Lytic and lysogenic life cycles of a temperate coliphage λ that infects Escherichia coli [44, 45]. cos—cohesive sites: the joining ends that circularise the linear phage λ DNA. Image courtesy of Springer Nature: https://www.nature.com/articles/nrg1089.

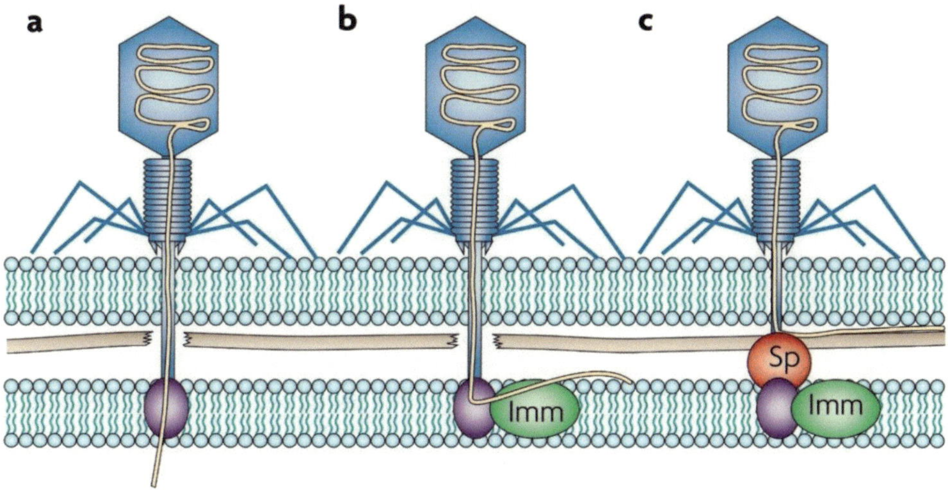

Figure 4.
Superinfection exclusion systems preventing phage DNA entry in Gram-negative Escherichia coli. (a). Standard T4 phage: upon attachment to phage-receptor on the surface of the host cell, an inner-membrane protein aids the translocation of phage DNA into the cell's cytoplasm. (b) Imm encoding phage T4: Imm protein directly blocks the translocation of the phage DNA into the cytoplasm of the cell. (c) Imm and Sp encoding phage T4: phage DNA is prevented from entering the cell's cytoplasm by Imm; and Sp protein prevents degradation of the peptidoglycan layer by inhibiting the activity of T4 lysozyme [28–31]. Image courtesy of Springer Nature: https://www.ncbi.nlm.nih.gov/pubmed/20348932.

5. Host strategies to cleave invading genomes and evolutionary tactics employed by phages to bypass these antiviral mechanisms

The evolution of bacterial genomes allowed bacteria to acquire vast mechanisms interfering with every step of phage infection. In a case where a phage succeeded to inject its viral nucleic acid into a host cell, bacteria possess a variety of nucleic acid degrading systems such as restriction-modification (R-M) systems and CRISPR/Cas that protect bacteria from the phage invasion.

5.1 Restriction-modification systems

It has been reported that R-M systems can significantly contribute to bacterial resistance to phages [49].

R-M systems incorporate activities of methyltransferases (MTases) that catalyse the transfer of a methyl group to DNA to protect self-genome from a restriction endonuclease (REase) cleavage and REases, which recognise and cut foreign unmethylated double-stranded DNA at specific recognition sites, commonly palindromic. To protect self-DNA from the degradation, methylases tag sequences recognised by the endonucleases with the methyl groups, whereas unmethylated phage (nonself) DNA is cleaved and degraded (**Figure 5**) [26, 27, 50–52].

R-M systems are diverse and ubiquitous among bacteria. There are four known types of R-M within bacterial genomes (**Figure 6**). Their classification is mainly based on R-M system subunit composition, sequence recognition, cleavage position, cofactor requirements and substrate specificity [26, 27, 50, 51].

Due to the diversity of R-M systems, phages acquired several active and passive strategies to bypass cleavage by REases. Passive mechanisms include reduction in restriction sites, modification and change of the orientation of restriction sites, whereas more specific, active mechanisms include masking of restriction sites,

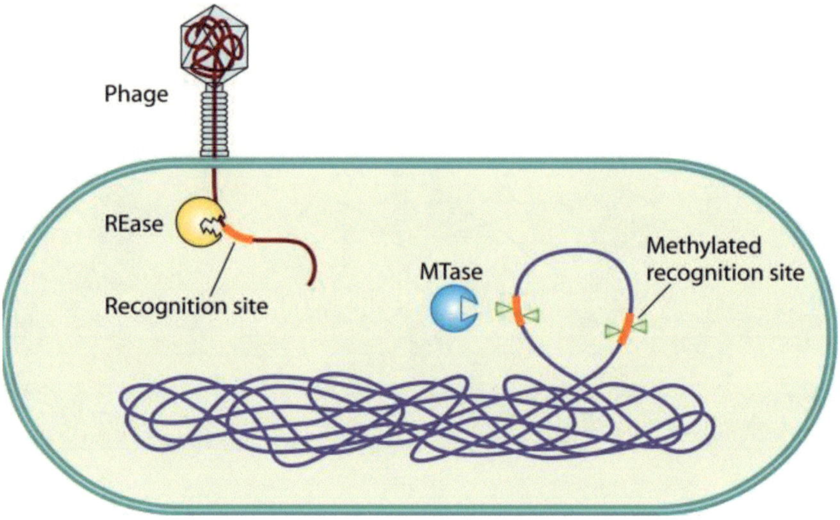

Figure 5.
General representation of the bacterial restriction-modification (R-M) systems providing a defence against invading phage genomes. R-M systems consist of two contrasting enzymatic activities: a restriction endonuclease (REase) and a methyltransferase. REase recognises and cuts nonself unmethylated double-stranded DNA at specific recognition sites, whereas MTase adds methyl groups to the same genomic recognition sites on the bacterial DNA to protect self-genome from REase cleavage [50, 51]. Image courtesy of: https://www.ncbi.nlm.nih.gov/pmc/articles/PMC3591985/.

stimulation of MTase activity on phage genome or degradation of an R-M system cofactor (**Figure 7**) [15–18].

Fewer restriction sites in the evading genome lead to the selective advantage of this phage as its DNA is less prone to cleavage and degradation by the host REase (**Figure 7a**). Also, some phages incorporate modified bases in their genomes that may lead to successful infection of the host cell as REase may not recognise the new sequences in the restriction sites. A decrease in the effective number of palindromic sites in DNA or change in the orientation of restriction-recognition sites can affect R-M targeting. Alternatively, the recognition sites within the viral genome can be too distant from each other to be recognised and cleaved by the REase [15–18, 53].

Interestingly, phage genome might be methylated by bacterial MTase upon successful injection into a host cell. Methylated recognition sites on viral genomes are therefore being protected from the cleavage and degradation by REase, leading to the initiation of the phage's lytic cycle. Viral progeny remains insensitive to this specific bacterial REase until it infects a bacterium that possesses a different type of REase, in which case the new progeny will become unmethylated again and will, therefore, be sensitive to the R-M system of the cognate bacterium [28–31].

The fate of the host cell chiefly confides in the levels of R-M gene expression and ultimate proportion of the R-M enzymes and their competition for the sites in the invading phage genome [52].

Furthermore, some phages encode their own MTase that is cooperative with the host REase, and thereby viral DNA cannot be recognised as nonself. Phages can also stimulate the activity of host modification enzymes that can rapidly methylate viral DNA, thus protecting it from the activity of REase.

Alternatively, phages can bypass R-M systems by masking restriction sites. For example (**Figure 7c**), coliphage P1, while injecting its DNA into a host cell, it also co-injects host-genome-binding proteins (DarA and DarB) that mask R-M recognition sites [53, 54].

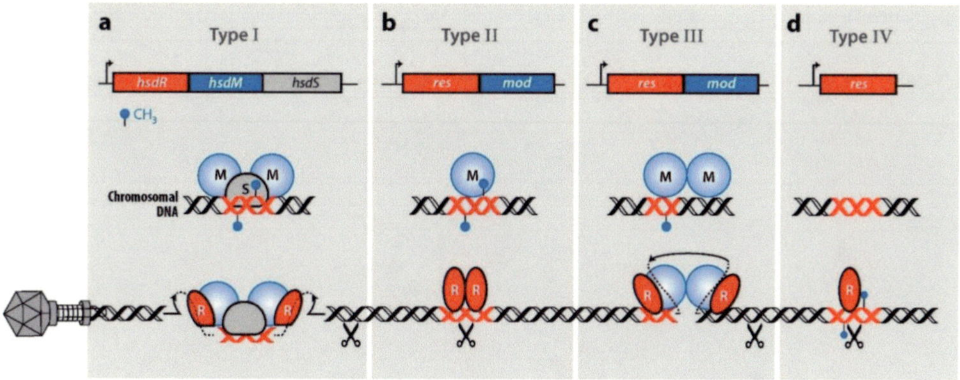

Figure 6.
Four distinct types of restriction-modification (R-M) systems. (a) Type I R-M system is composed of three subunits forming a complex: hsdR (restriction), hsdM (modification) and hsdS (specificity subunit that binds to an asymmetrical DNA sequence and determines the specificity of restriction and methylation). Two hsdM subunits and one hsdS subunit are involved in methylation of self-DNA. On the other hand, two complexes of hsdR, hsdM and hsdS (where each complex consists of two hsdR, two hsdM and one hsdS subunit) bind to the unmethylated recognition sites on phage DNA and cleave the DNA at random, far from their recognition sequences. Both reactions—methylation and cleavage—require ATP. (b) Type II R-M system is composed of two distinct enzymes: palindromic sequence methylating methyltransferase (mod) and endonuclease (res) that cleave unmethylated palindromic sequences close to or within the recognition sequence. (c) Type III R-M system is formed of methyltransferase (mod) and endonuclease (res) that form a complex. Methyltransferase transfers methyl group to one strand on the DNA, whereas two methyltransferases (endonuclease complexes) act together to bind to the complementary unmethylated recognition sites to cleave the DNA 24–26 bp away from the recognition site. (d) Type IV R-M system contains only endonuclease (res) that recognises methylated or modified DNA. Cleavage occurs within or away from the recognition sequences [26, 27, 50, 51]. Image courtesy of: https://www.annualreviews.org/doi/abs/10.1146/annurev-virology-031413-085500?journalCode=virology.

As shown on an example of a Coliphage T7 (**Figure 7d**), some phages code for proteins that directly inhibit REase. Coliphage T7 possesses proteins that can mimic the DNA backbone. Ocr, a protein expressed by Coliphage T7, directly blocks the active site of some REases by mimicking 24 bp of bent B-form DNA, and it has a high affinity for the EcoKI REase component, thereby interfering with R-M system [53].

Lastly, phage-bacteria arm race allowed phages to gain capabilities of degrading necessary cofactors of R-M systems. For instance, coliphage T3 encodes S-adenosyl-l-methionine hydrolase that destroys an essential host R-M cofactor (the S-adenosyl-l-methionine). The removal of this necessary co-factor will lead to the inhibition of the REase, thereby successfully infecting the host cell [15–18].

5.2 CRISPR/Cas system

CRISPR along with CRISPR-associated (Cas) proteins is the type of adaptive heritable 'immunity' of bacteria, thus very specific and effective; and it is prevalent within the bacterial domain [55]. The CRISPR are DNA loci consisting of short palindromic repeats (identical in length and sequence), interspaced by segments of DNA sequences (spacer DNA) derived from previous exposures to phages. The spacer DNA sequences act as a 'memory', allowing bacteria to recognise and destroy specific phages in a subsequent infection. Genes encoding Cas proteins are adjacent to CRISPR loci [56].

Although some studies have suggested that CRISPRs can be used for pathogen subtyping [57], it has been found that CRISPR typing is not useful for the epidemiological surveillance and outbreak investigation of *Salmonella typhimurium* [58].

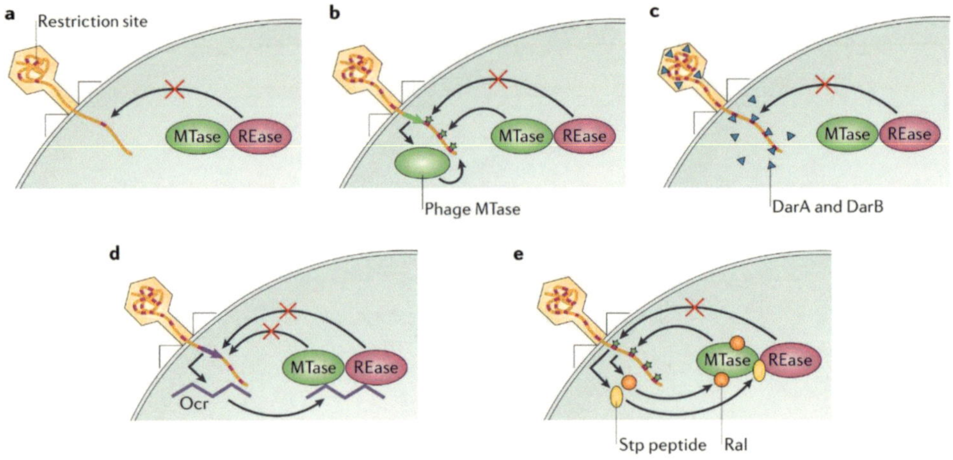

Figure 7.
Phage's passive and active strategies to bypass restriction-modification (R-M) systems. (a) Phages that possess fewer restriction sites in their genome are less prone to DNA cleavage by the host restriction endonuclease (REase). (b) Occasionally phage DNA might be modified by bacterial methyltransferase (MTase) upon successful injection into a host cell. Methylated recognition sites on viral DNA are, therefore, being protected from the cleavage and degradation by REase, leading to the initiation of the phage's lytic cycle. In addition, some phages encode their own MTase that is cooperative with the host REase; thus viral DNA cannot be recognised as nonself. (c) Some phages, for example, coliphage P1, while injecting its DNA into a host cell, it also co-injects host-genome-binding proteins (DarA and DarB) that mask R-M recognition sites. (d) Phages such as Coliphage T7 possess proteins that can mimic the DNA backbone. Ocr, a protein expressed by Coliphage T7, mimics the DNA phosphate backbone and has a high affinity for the EcoKI REase component, thereby interfering with R-M system. (e) In addition, some phages (e.g. Ral protein of Coliphage λ) can also stimulate activity of the bacterial modification enzyme in order to protect own DNA from the recognition by the bacterial REase as nonself. The peptide Stp encoded by Coliphage T4 can as well disrupt the structural conformation of the REase-MTase complex [15–18]. Image courtesy of: https://www.nature.com/articles/nrmicro3096.

The CRISPR/Cas phage resistance is mediated in three-step stages: adaptation (acquisition), where spacer phage-derived DNA sequences are incorporated into the CRISPR/Cas system; expression, where *cas* gene expression and CRISPR transcription lead to pre-CRISPR RNA (pre-crRNA) that is then processed into CRISPR RNA (crRNA); and interference, during which the crRNA guides Cas proteins to the target (subsequently invading DNA) for the degradation. The cleavage of the target (proto-spacer) depends on the recognition of complementary sequences in spacer and protospacer [59, 60].

CRISPR/Cas systems have been classified into three major types: Types I, II and III, which are further divided into subtypes that require different types of Cas proteins. Although the CRISPR/Cas array is diverse among the bacteria and it is continuously co-evolving in response to the host-phage interactions, the defence activity in all three types of the CRISPR is comparable [21–23] **Figure 8** illustrates the defence mechanisms in three distinct CRISPR/Cas arrays.

The Type II, CRISPR/Cas9, which was first identified in *Streptococcus pyogenes, gained considerable interest within scientific studies as a precise genome editing tool. CRISPR/Cas9 system is unique; a single Cas 9 protein (in addition to prevalent Cas 1 and Cas 2) is involved in the processing of* crRNA and destruction of the target viral DNA [56, 61].

In the adaptation stage, phage-derived protospacer (snippet of DNA from the invading phage) is incorporated into the bacterial genome at the leader end of the CRISPR loci. In expression phase, the *Cas9* gene expresses Cas9 protein possessing DNA cleaving HNH and RuvC-like nuclease domains; CRISPR locus is then transcribed and processed into mature crRNA. Finally, in interference step, the complex consisting of Cas9, crRNA and separate trans-activating crRNA (tracrRNA) cleave 20 base pairs crRNA-complementary target sequence that is adjacent to the protospacer adjacent motif (PAM) [62].

To bypass CRISPR/Cas that has an incredibly dynamic rate of evolution, phages acquired array of strategies to succeed in propagation; this includes mutations in the protospacers or in the PAM sequences and expression of anti-CRISPR proteins, and even some phages encode their own functional CRISPR/Cas systems [15–18, 63].

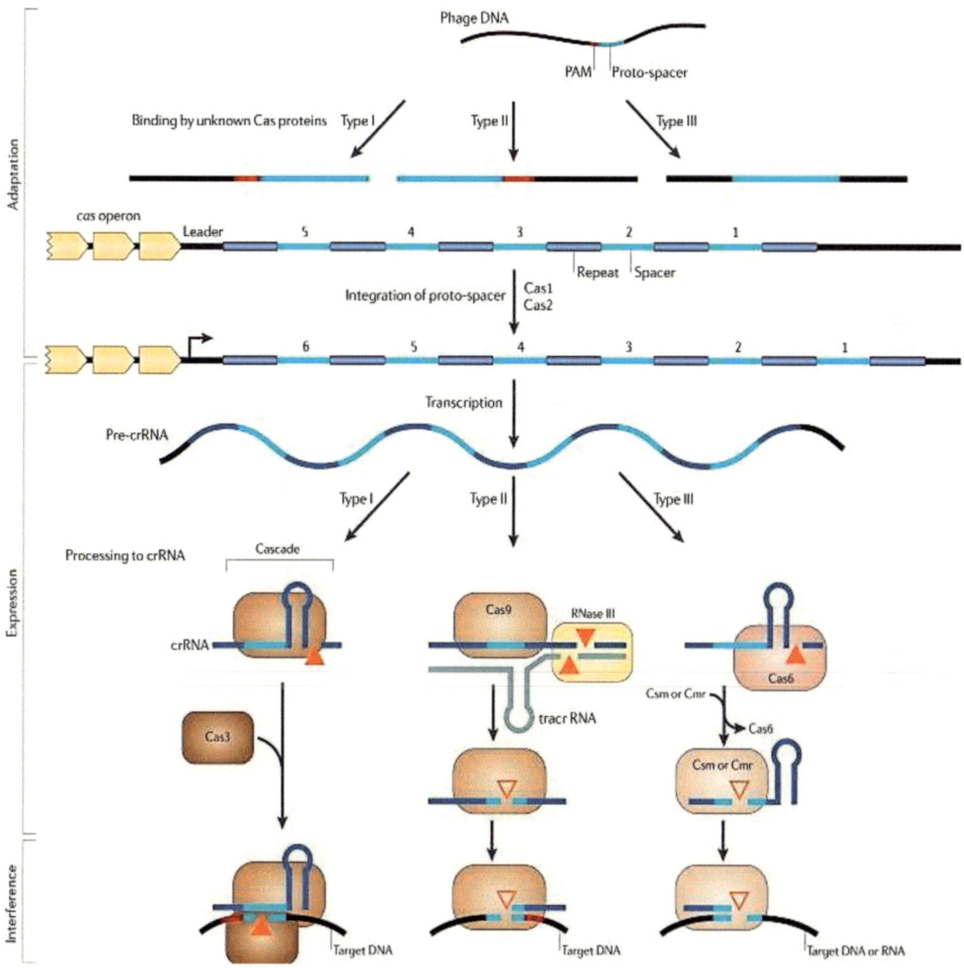

Figure 8.
Image showing mechanisms of adaptation, expression and interference in three different types of CRISPR/Cas arrays. Type I and Type II CRISPR/Cas arrays rely on the protospacer adjacent motif (PAM), contained within phage nucleic acid, to 'select' the phage-derived protospacer. Next steps in the adaptation stage are similar in all three types; protospacer is incorporated by Cas 1 and Cas2 proteins into the bacterial genome at the leader end of the CRISPR loci to form a new spacer. In expression step, CRISPR loci are transcribed into pre-crRNA. The crRNA processing and interference stage is distinct in each type of the CRISPR/Cas system. In Type I, the multisubunit CRISPR-associated complex for antiviral defence (CASCADE) binds crRNA to locate the target, and with the presence of Cas3 protein, the invading target genome is degraded whereas in Type II, Cas9 protein is essential in the processing of the crRNA. TracrRNA recognises and attaches to the complementary sequences on the repeat region that is then cut by RNase III in the presence of Cas9. Lastly, in Type III, processing of pre-crRNA into crRNA is dependent upon the activity of Cas6. Mature crRNA associated with Csm/Cmr complex targets foreign DNA or RNA for the degradation [21–23]. Image courtesy of: https://www.nature.com/articles/nrmicro2577.

Phages can evade interference step of Type I and Type II CRISPR/Cas system by a single point mutation or deletion in their protospacer region or in the PAM sequence (**Figure 9**). Phages with single-nucleotide substitutions or deletions positioned close to PAM sequence can bypass the CRISPR/Cas activity and complete their lytic cycles; in contrast, phages with multiple mutations at PAM-distal protospacer positions do not [15–18, 28–31].

In some circumstances, however, although the phage successfully evades CRISPR/Cas interference, the host cell may survive by the acquisition of new spacer sequences (derived from invading phage) into their own CRISPR/Cas system. This new spacer provides the bacterium with an accelerated spectrum of phage resistance [15–18].

Prophages integrated within *Pseudomonas aeruginosa* possess genes that encode anti-CRISPR proteins directly suppressing CRISPR/Cas-mediated degradation of the phage genome (**Figure 10**). According to Wiedenheft [64], these proteins might interrupt CRISPR RNA processing by preventing mature crRNA from binging to the crRNA-guide complex or by preventing the assembled crRNA-guided complex from interacting with target substrates through binding to it [64].

Prophages do not only contribute to bacterial resistance to invading phages, they can also encode proteins that contribute to bacterial virulence and antimicrobial resistance [58, 66].

Bacteria can also resist phages by possessing phage-inducible chromosomal islands (PICI) which prevent phage replication. Nevertheless, phages evolved their genomes to overcome this very specific antiviral strategy. For example, *Vibrio cholerae* ICP1 phages possess their own CRISPR/Cas systems that inactivate PICI-like elements (PLE) in *Vibrio cholerae* (**Figure 11**). Studies conducted by Naser et al. [67] have shown that phage CRISPR arrays have evolved by the acquisition of new spacers targeting diverse regions of PLEs carried by *Vibrio cholerae* strains. Furthermore, the addition of the new spacers within phage CRISPR/Cas loci enables the phages to expand their ability to counter PLE-mediated phage defence of diverse *Vibrio cholerae* strains [67].

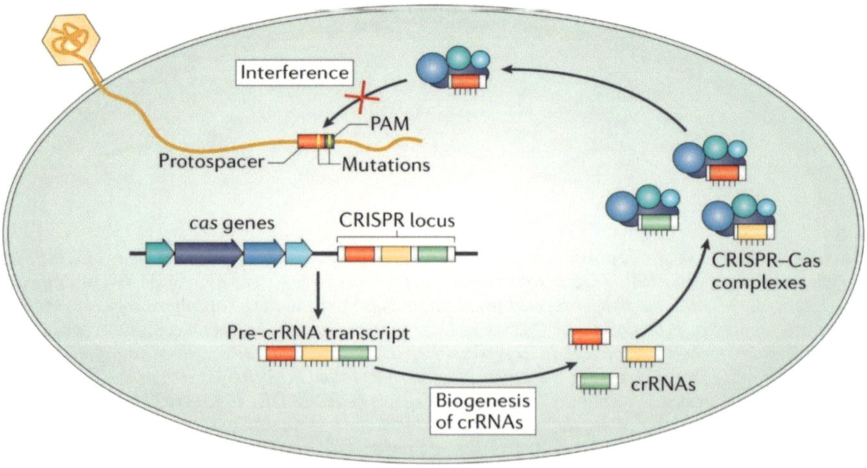

Figure 9.
Evasion by mutation. Mutations in the phage protospacers or in the PAM sequences allow the phage to escape interference step of the CRISPR/Cas system that would lead to the degradation of the phage genome [15–18]. Adapted image courtesy of: https://www.nature.com/articles/nrmicro3096.

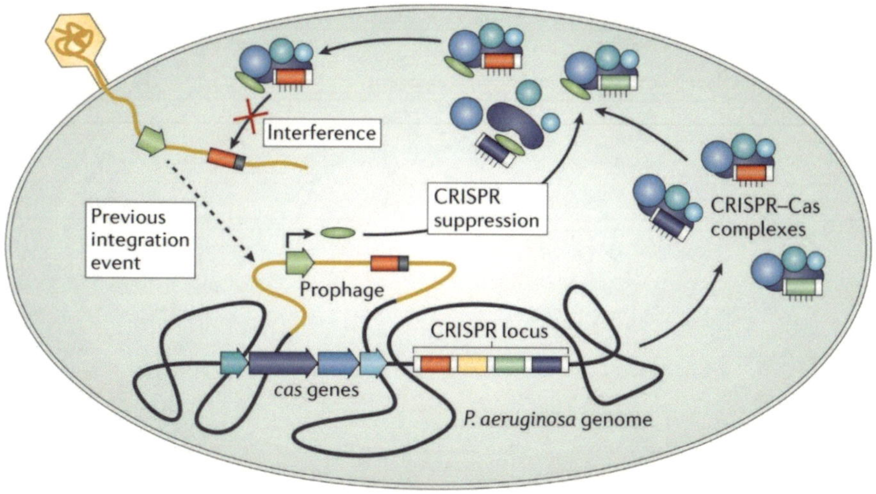

Figure 10.
Anti-CRISPR proteins expressed against CRISPR subtype I-F systems. Temperate phages such as Pseudomonas aeruginosa possess genes encoding anti-CRISPR proteins that directly interfere with the bacterial CRISPR/Cas system [15–18]. Adapted image courtesy of: https://www.nature.com/articles/nrmicro3096.

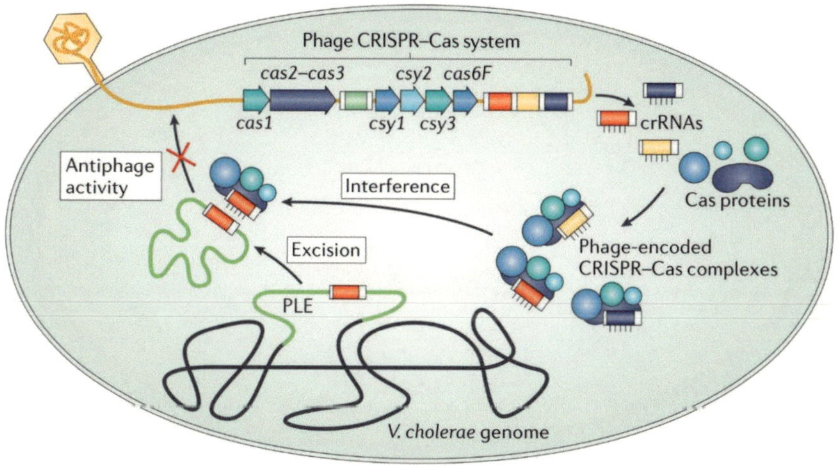

Figure 11.
Phage-encoded CRISPR/Cas systems in Vibrio cholerae ICP1 phages. Upon adsorption and injection of viral genome into a host cell, phage crRNAs and CRISPR/Cas complexes are expressed and target phage-inducible chromosomal island (PICI) in the host genome; in the Vibrio cholerae, they are termed as PICI-like elements (PLE). If the spacers within phage CRISPR locus are complementary to the bacterial PLE, the CRISPR machinery is then able to specifically target this genetic element and inactivate it, leading to the viral propagation. However, in the absence of such targeting, phage CRISPR/Cas system can acquire new spacers to evolve rapidly and ensure effective targeting of the PLE to restore phage replication [15–18, 65]. Adapted image courtesy of: https://www.nature.com/articles/nrmicro3096.

6. Overcoming host abortive infection systems: toxin-antitoxin

Abortive infection (Abi) systems promote cell death of the phage-infected bacteria, inhibiting phage replication and providing protection for bacterial populations [68].

Abi systems require both toxins and antagonistic antitoxins. Antitoxins are proteins or RNAs that protect bacterial cell from the activity of toxins in a typical cell life cycle, whereas toxins are the proteins encoded in toxin-antitoxin locus that

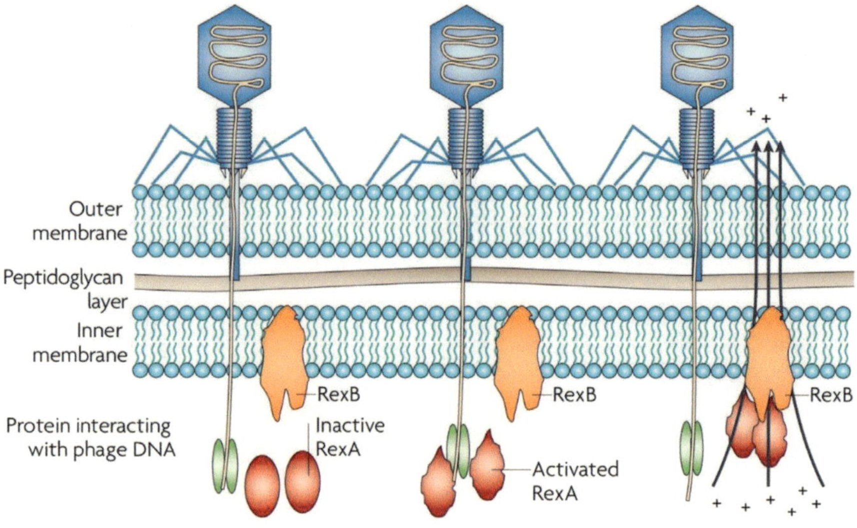

Figure 12.
Abortive infection (Abi) systems in Escherichia coli. The Rex system is a two-component Abi system. A phage protein-DNA complex (formed during phage replication) activates the sensor protein RexA, which in turn activates RexB. RexB is an ion channel that causes depolarisation of the bacterial membrane leading to cell death [28–31]. Image courtesy of Springer Nature: https://www.ncbi.nlm.nih.gov/pubmed/20348932.

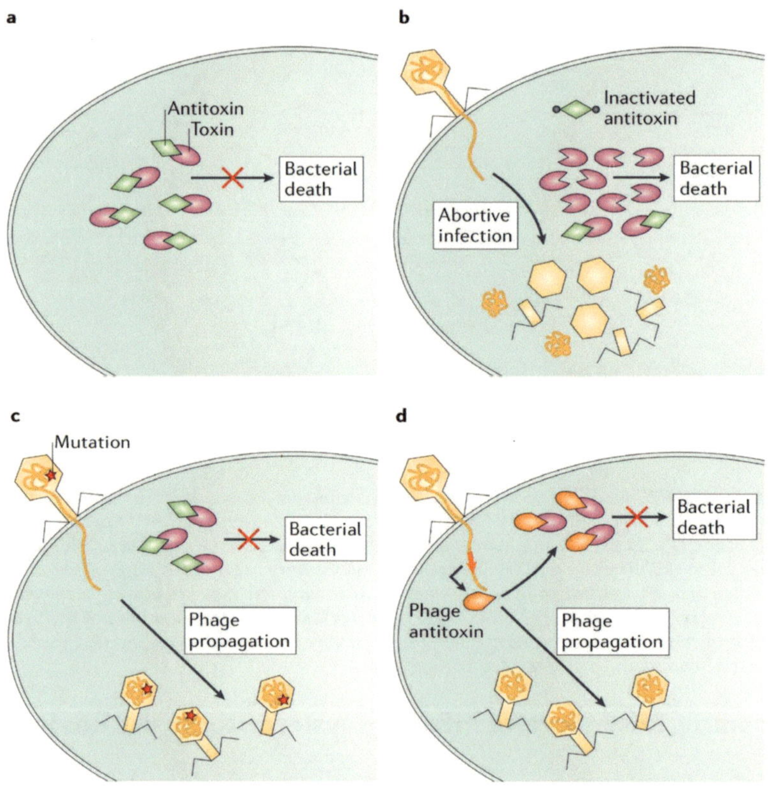

Figure 13.
Escaping abortive infection mechanisms. (a) In a typical cell life cycle, antitoxins protect bacterial cell from the activity of toxins. (b) During phage infection, the expression of antitoxin encoding gene is suppressed, leading to the lethal activation of the toxin. (c) Mutations in certain phage genes can lead to escaping Abi systems activity, thereby a successful viral propagation without killing the host cell. (d) Some phages encode molecules that functionally replace the bacterial antitoxins, thus suppressing toxin activity and avoiding host cell death [15–18]. Image courtesy of: https://www.nature.com/articles/nrmicro3096.

disrupt cellular metabolism (translation, replication and cell wall formation), causing cell death. During an infection, the expression of the antitoxin encoding gene is suppressed, leading to the lethal activation of the toxin [69]. **Figure 12** illustrates the mechanism of Abi systems in *Escherichia coli* [70].

Interestingly, phages evolved an array of tactics to circumvent Abi systems. This includes mutations in specific phage genes and encoding own antitoxin molecules that suppresses bacterial toxin [15–18]. **Figure 13** provides a broad overview of the strategies employed by the phages to by-pass Abi systems.

Bacteria-phage interaction is therefore very complex, and it is crucial to understand the molecular basis of this interaction and how bacteria and phages 'fight' each other. It has been reported that Anderson Phage Typing System of *Salmonella* Typhimurium can provide a valuable model system for study of phage-host interaction [71].

7. The potential application of phages as antibacterial therapeutics

The rapid emergence and dissemination of MDR bacteria seriously threaten global public health, as, without effective antibiotics, prevention and treatment of both community- and hospital-acquired infections may become unsuccessful and lead to widespread outbreaks.

Carbapenems and colistin are antibiotics of last resort, generally reserved to treat bacteria which are resistant to all other antibiotics. Until not long ago, colistin resistance was only described as chromosomal, however, in 2016 Liu et al. reported the emergence of the first plasmid-mediated colistin resistance mechanism, MCR-1, in Enterobacteriaceae [72]. Furthermore, the increasing occurrence of colistin resistance among carbapenem-resistant Enterobacteriaceae has also been reported [73]. This is of significant concern as infections caused by colistin and carbapenem-resistant bacteria are very challenging to treat and control, as the treatment options are greatly limited or non-existent. Thus, the discovery and development of alternative antimicrobial therapeutics are the highest priorities of modern medicine and biotechnology.

Phages should be considered as great potential tools in MDR pathogens as they are species-specific (specificity prevents damage of normal microbiota), thus harmless to human; they have fast replication rate at the site of infection, and their short genomes can allow to further understand various molecular mechanisms implied to 'fight' bacteria. In addition, this understanding can enable scientists to 'manipulate' viral genomes and engineer a synthetic phage that combines the antibacterial characteristics of multiple phages into a single genome.

The escalating need for new antimicrobial agents attracted new attention in modern medicine, proposing several potential applications of phages as antibacterial therapeutics including phage therapy, phage lysins and genetically-engineered phages.

7.1 Phage therapy

Phage therapy utilises strictly lytic phages that have bactericidal effect. As phages are host-specific, 'phage cocktails' containing multiple phages can broaden range of target cells. Nevertheless, selection of suitable phages is at the paramount to the successful elimination of clinically important pathogens, and it includes avoidance of adverse effects, such as anaphylaxis (adverse immune reaction) [74].

7.2 Phage-derived enzymes: lysins

In order to hydrolyse and degrade the bacterial cell wall, phages possess lysins.

The spectrum of efficiency of natural lysins (derived from naturally occurring phages) is generally limited to Gram-positive bacteria; however, recombinant lysins have shown an ability to destabilise the outer membrane of Gram-negative bacteria and ultimately lead to rapid death of the target bacteria [74].

7.3 Bioengineered phages

Bioengineered phages have the potential to solve inherent limitations of natural phages such as narrow host range and evolution of resistance. Various genetic engineering methods have been proposed to design phages with extended antimicrobial properties such as homologous recombination, phage recombineering of electroporated DNA, yeast-based platform, Gibson assembly and CRISPR/Cas genome editing [75].

Engineering of synthetic phages could be tailored to enhance the antibiotic activity, to reverse antibiotic resistance or to create sequence-specific antimicrobials [74].

8. Conclusions

The antagonistic host-phage relationship has led to the evolution of exceptionally disperse phage-resistance mechanisms in the bacterial domain, including inhibition of phage adsorption, prevention of nucleic acid entry, Superinfection exclusion, cutting phage nucleic acids via restriction-modification systems and CRISPR, as well as abortive infection.

Evolvement of these mechanisms has been induced by constant parallel co-evolution of phages as they attempt to coexist. To survive, phages acquired diverse counterstrategies to circumvent bacterial anti-phage mechanisms such as adaptations to new receptors, digging for receptors and masking and modification of restriction sites and point mutations in specific genes and genome rearrangements that allow phages to evade bacterial antiviral systems such as CRISPR/Cas arrays, as well as mutations in specific genes to bypass abortive infection system. Conclusively, the co-evolving genetic variations and counter-adaptations, in both bacteria and phages, drive the evolutionary bacteria-host arm race.

Besides, accumulating evidence shows that phages contribute to the antimicrobial resistance through horizontal gene transfer mechanisms. Indeed, many bacterial strains have become insensitive to the conventional antibiotics, posing a growing threat to human; and although in the past, western counties withdrew phage therapy in response to the discovery of therapeutic antibiotics, now, phage therapy regains an interest within the research community. There are apparent advantages of phage therapy, such as specificity, meaning only target bacteria would encounter lysis, but not healthy microbiota inhabiting human's system. Additionally, 'phage cocktails', containing multiple bacteria-specific phages, could overcome the issue of phage-resistance as phages do adapt to these resistance mechanisms. However, 'phage cocktails' would require large numbers of phages that would have to be grown inside pathogenic bacteria in the laboratory, putting laboratory staff and the environment at risk.

Alternatively, building up the understanding of host-phage interactions and 'the war between bacteria and phages' could potentially lead to defeating antimicrobial resistance by designing synthetic phages that can overcome the limitations of phage therapy.

Acknowledgements

Dr Manal Mohammed is funded by a Quinton Hogg start-up award, University of Westminster.

Abbreviations

Abi	abortive infection
CPS	capsular polysaccharides
CRISPR	clustered regularly interspaced short palindromic repeats
crRNA	crispr RNA
DGR	diversity-generating retroelement
DNA	deoxyribonucleic acid
MDR	multidrug-resistant
MeOPN	O-methyl phosphoramidate
MTase	methyltransferase
PAM	protospacer adjacent motif
PICI	phage-inducible chromosomal island
PLE	PICI-like element
RBP	receptor-binding protein
REase	restriction endonuclease
R-M	restriction-modification
RNA	ribonucleic acid
Sie	superinfection exclusion
tracrRNA	trans-activating crRNA

Author details

Beata Orzechowska and Manal Mohammed*
School of Life Sciences, College of Liberal Arts and Sciences University of Westminster, London, UK

*Address all correspondence to: m.mohammed@westminster.ac.uk

References

[1] Public Health England. Health Matters: Antimicrobial Resistance. Government Digital Service. 2018. Available from: https://www.gov.uk/government/publications/health-matters-antimicrobial-resistance/health-matters-antimicrobial-resistance [Accessed: 10 July 2018]

[2] Slama TG. Gram-negative antibiotic resistance: There is a price to pay. Critical Care. 2008;**12**:S4

[3] Herelle FD. The Bacteriophage: Its Rôle in Immunity. English ed. Baltimore: Williams & Wilkins; 1922

[4] Sulakvelidze A, Alavidze Z, Morris JG. Bacteriophage therapy. Antimicrobial Agents and Chemotherapy. 2001;**45**(3):649

[5] Fauconnier A. Regulating phage therapy. EMBO Reports. 2017;**18**(2):198-200

[6] Baharuddin A et al. Revitalizing phage therapy in combating multi-drug resistant bacteria. Haya: The Saudi Journal of Life Sciences. 2017;**2**(4):122-130

[7] Bikard D, Marraffini LA. Innate and adaptive immunity in bacteria: Mechanisms of programmed genetic variation to fight bacteriophages. Current Opinion in Immunology. 2012;**24**(1):15-20

[8] Breitbart M, Rohwer F. Here a virus, there a virus, everywhere the same virus? Trends in Microbiology. 2005;**13**(6):278-284

[9] Muhammad Abu BS et al. Bacteria vs. bacteriophages: Parallel evolution of immune arsenals. Frontiers in Microbiology. 2016;7:1292

[10] Chao L, Levin BR, Stewart FM. A complex community in a simple habitat: An experimental study with bacteria and phage. Ecology. 1977;**58**(2):369-378

[11] Bohannan B, Lenski R. Linking genetic change to community evolution: Insights from studies of bacteria and bacteriophage. Ecology Letters. 2000;**3**(4):362-377

[12] Chibani-Chennoufi S, Bruttin A, Dillmann M, Brussow H. Phage-host interaction: An ecological perspective. The Journal of Bacteriology. 2004;**186**(12):3677

[13] Van Valen L. A new evolutionary law. Evolutionary Theory. 1973;**1**(1):1-30

[14] Betts A et al. Contrasted coevolutionary dynamics between a bacterial pathogen and its bacteriophages. Proceedings of the National Academy of Sciences. 2014;**111**(30):11109

[15] Samson JE et al. Passive and active strategies to avoid restriction-modification systems [image]. 2013. Available from: https://www.nature.com/articles/nrmicro3096 [Accessed: 9 July 2018]

[16] Samson JE et al. Phage strategies to by-pass CRISPR-Cas systems [image]. 2013. Available from: https://www.nature.com/articles/nrmicro3096 [Accessed: 9 July 2018]

[17] Samson JE et al. Phage strategies to by-pass toxin–antitoxin systems [image]. 2013. Available from: https://www.nature.com/articles/nrmicro3096 [Accessed: 11 July 2018]

[18] Samson JE et al. Revenge of the phages: Defeating bacterial defences. Nature Reviews Microbiology. 2013;**11**(10):675

[19] Hyman P, Abedon ST. Bacteriophage host range and bacterial resistance—

chapter 7. Advances in Applied Microbiology. 2010;**70**:217-248

[20] Doron S et al. Systematic discovery of antiphage defense systems in the microbial pangenome. Science (New York, N.Y.). 2018;**359**(6379):eaar4120

[21] Makarova KS et al. Evolution and classification of the CRISPR-Cas systems. Nature Reviews. Microbiology. 2011;**9**(6):467

[22] Makarova KS et al. Defense Islands in bacterial and Archaeal genomes and prediction of novel defense systems. Journal of Bacteriology. 2011;**193**(21):6039-6056

[23] Makarova KS et al. The three stages of CRISPR–Cas action [image]. 2011. Available from: https://www.nature.com/articles/nrmicro2577 [Accessed: 8 July 2018]

[24] Vale PF, Little TJ. CRISPR-mediated phage resistance and the ghost of coevolution past. Proceedings of the Royal Society B. 2010;**277**(1691):2097-2103

[25] Etz H et al. Bacterial phage receptors, versatile tools for display of polypeptides on the cell surface. The Journal of Bacteriology. 2001;**183**(23):6924

[26] Dy RL et al. The mechanisms for methylation and restriction for the four types of restriction-modification (RM) systems [image]. 2014. Available from: https://www.annualreviews.org/doi/abs/10.1146/annurev-virology-031413-085500?journalCode=virology [Accessed: 8 July 2018]

[27] Dy RL et al. Remarkable mechanisms in microbes to resist phage infections. Annual Review of Virology. 2014;**1**:307-331

[28] Labrie SJ, Samson JE, Moineau S. Bacteriophage resistance mechanisms. Nature Reviews Microbiology. 2010;**8**(5):317

[29] Labrie SJ et al. Blocking phage DNA entry into the bacterial cell [image]. 2010. Available from: https://www.ncbi.nlm.nih.gov/pubmed/20348932 [Accessed: 9 July 2018]

[30] Labrie SJ et al. Different strategies used by bacteria to block phage adsorption [image]. 2010. Available from: https://www.ncbi.nlm.nih.gov/pubmed/20348932 [Accessed: 9 July 2018]

[31] Labrie SJ et al. The mode of action of the rex system [image]. 2010. Available from: https://www.ncbi.nlm.nih.gov/pubmed/20348932 [Accessed: 6 July 2018]

[32] Golais F, Hollý J, Vítkovská J. Coevolution of bacteria and their viruses. Folia Microbiologica. 2013;**58**(3):177-186

[33] Ahmad S et al. An overview on phase variation, mechanisms and roles in bacterial adaptation. JPMA. The Journal of the Pakistan Medical Association. 2017;**67**(2):285-291

[34] Guo H et al. Diversity-generating retroelements in phage and bacterial genomes. Microbiology Spectrum. 2014;**2**(6):2014

[35] Sørensen MC et al. The F336 bacteriophage recognizes the capsular phosphoramidate modification of Campylobacter jejuni NCTC11168. Journal of Bacteriology. 2011;**193**:6742-6749. DOI: 10.1128/JB.05276-11

[36] Sørensen MC et al. Phase variable expression of capsular polysaccharide modifications allows Campylobacter jejuni to avoid bacteriophage infection in chickens. Frontiers in Cellular and Infection Microbiology. 2012;**2**:11. DOI: 10.3389/fcimb.2012.00011

[37] Dai W et al. Three-dimensional structure of tropism-switching Bordetella bacteriophage. Proceedings of the National Academy of Sciences of the United States of America. 2010;**107**(9):4347

[38] Nordstrom K, Forsgren A. Effect of protein a on adsorption of bacteriophages to *Staphylococcus aureus*. The Journal of Virology. 1974;**14**(2):198

[39] Nwodo U, Green E, Okoh A. Bacterial Exopolysaccharides: Functionality and Prospects. Basel: MDPI AG; 2012

[40] Looijesteijn PJ et al. Physiological function of exopolysaccharides produced by *Lactococcus lactis*. International Journal of Food Microbiology. 2001;**64**(1):71-80

[41] Sutherland IW. Polysaccharide lyases. FEMS Microbiology Reviews. 1995;**16**(4):323

[42] Letarov A, Kulikov E. Adsorption of bacteriophages on bacterial cells. Biochemistry (Moscow). 2017;**82**(13):1632-1658

[43] Orlova EV. How viruses infect bacteria? EMBO Journal. 2009;**28**(7):797-798

[44] Campbell A. Life cycle of the typical temperate phage coliphage-λ [image]. 2003. Available from: https://www.nature.com/articles/nrg1089 [Accessed: 7 July 2018]

[45] Campbell A. The future of bacteriophage biology. Nature Reviews Genetics. 2003;**4**(6):471

[46] St-Pierrea F, Endy D. Determination of cell fate selection during phage lambda infection. Proceedings of the National Academy of Sciences of the United States of America. 2008;**105**(52):20705-20710

[47] Folimonova SY. Superinfection exclusion is an active virus-controlled function that requires a specific viral protein. Journal of Virology. 2012;**86**(10):5554

[48] Hofer B, Ruge M, Dreiseikelmann B. The superinfection exclusion gene (sieA) of bacteriophage P22: Identification and overexpression of the gene and localization of the gene product. The Journal of Bacteriology. 1995;**177**(11):3080

[49] Mohammed M, Cormican M. Whole genome sequencing provides possible explanations for the difference in phage susceptibility among two Salmonella Typhimurium phage types (DT8 and DT30) associated with a single foodborne outbreak. BMC Research Notes. 2015;**8**(1):728

[50] Vasu K, Nagaraja V. Diverse Functions of Restriction-Modification Systems in Addition to Cellular Defense. Washington: American Society for Microbiology; 2013

[51] Vasu K, Nagaraja V. Restriction-modification (R-M) systems as defense mechanisms [image]. 2013. Available from: https://www.ncbi.nlm.nih.gov/pmc/articles/PMC3591985/ [Accessed: 5 July 2018]

[52] Enikeeva FN et al. Restriction-modification systems and bacteriophage invasion: Who wins? Journal of Theoretical Biology. 2010;**266**(4):550-559

[53] Stern A, Sorek R. The phage-host arms race: Shaping the evolution of microbes. BioEssays. 2011;**33**(1):43-51

[54] Tock MR, Dryden DT. The biology of restriction and anti-restriction. Current Opinion in Microbiology. 2005;**8**(4):466-472

[55] Marchfelder A. Special focus CRISPR-Cas. RNA Biology. 2013;**10**(5):655-658

[56] Heler R et al. Cas9 specifies functional viral targets during CRISPR-Cas adaptation. Nature. 2015;**519**(7542):202L

[57] Fabre L et al. CRISPR typing and subtyping for improved laboratory surveillance of salmonella infections (CRISPR polymorphisms in salmonella). PLoS One. 2012;7(5):e36995

[58] Mohammed M et al. The invasome of Salmonella Dublin as revealed by whole genome sequencing. BMC Infectious Diseases. 2017;**17**(1):544

[59] Koonin EV, Makarova KS. CRISPR-Cas: Evolution of an RNA-based adaptive immunity system in prokaryotes. RNA Biology. 2013, 2013;**10**(5):679-686

[60] Leon LM, Mendoza SD, Bondy-Denomy J. How bacteria control the CRISPR-Cas arsenal. Current Opinion in Microbiology. 2018;**42**:87-95

[61] Ka D et al. Crystal structure of streptococcus pyogenes Cas1 and its interaction with Csn2 in the type II CRISPR-Cas system. Structure. 2016;**24**(1):70-79

[62] Carroll D. Staying on target with CRISPR-Cas. Nature Biotechnology. 2013;**31**(9):807-809

[63] Mohanraju P et al. Diverse evolutionary roots and mechanistic variations of the CRISPR-Cas systems. Science. 2016;**353**(6299)

[64] Wiedenheft B. In defense of phage: Viral suppressors of CRISPR-mediated adaptive immunity in bacteria. RNA Biology. 2013;**10**(5):886

[65] Seed K et al. A bacteriophage encodes its own CRISPR/Cas adaptive response to evade host innate immunity. Nature. 2013;**494**(7438):489-491

[66] Mohammed M, Cormican M. Whole genome sequencing provides insights into the genetic determinants of invasiveness in Salmonella Dublin. Epidemiology and Infection. 2016;**144**(11):2430-2439

[67] Naser I et al. Analysis of the CRISPR-Cas system in bacteriophages active on epidemic strains of *Vibrio cholerae* in Bangladesh. Scientific Reports. 2017;**7**(1):14880

[68] Chopin M, Chopin A, Bidnenko E. Phage abortive infection in lactococci: Variations on a theme. Current Opinion in Microbiology. 2005;**8**(4):473-479

[69] Fineran PC et al. The phage abortive infection system, ToxIN, functions as a protein-RNA toxin-antitoxin pair. Proceedings of the National Academy of Sciences of the United States of America. 2009;**106**(3):894

[70] Refardt D, Kümmerli R. Defying bacteriophages: Contrasting altruistic with individual-based resistance mechanisms in *Escherichia coli*. Communicative & Integrative Biology. 2013;**6**(5):e25159

[71] Mohammed M. Who fights whom? Understanding the complex dynamics of bacteria-phage interaction using Anderson phage typing system. Journal of Infectious Diseases & Therapy. 2018;**6**(3):367

[72] Liu Y et al. Emergence of plasmid-mediated colistin resistance mechanism MCR-1 in animals and human beings in China: A microbiological and molecular biological study. The Lancet Infectious Diseases. 2016;**16**(2):161-168

[73] European Centre for Disease Prevention and Control. Expert consensus protocol on colistin resistance detection and characterisation for the survey of carbapenem- and/or colistin-resistant Enterobacteriaceae. 2019. Available from: https://ecdc.europa.eu/en/publications-data/expert-consensus-protocol-colistin-resistance-detection-and-characterisation [Accessed: 13 May 2019]

[74] Kakasis A, Panitsa G. Bacteriophage therapy as an alternative treatment for human infections. A comprehensive review. International Journal of Antimicrobial Agents. 2019;**53**(1):16-21

[75] Monteiro R et al. Phage therapy: Going temperate? Trends in Microbiology. 2019;**27**(4):368-378. DOI: 10.1016/j.tim.2018.10.008

Chapter 7

Mycelium Differentiation and Development of *Streptomyces* in Liquid Nonsporulating Cultures: Programmed Cell Death, Differentiation, and Lysis Condition Secondary Metabolite Production

Angel Manteca, Beatriz Rioseras, Nathaly González-Quiñónez, Gemma Fernández-García and Paula Yagüe

Abstract

Streptomycetes are mycelium-forming sporulating bacteria that produce two thirds of clinically relevant secondary metabolites. Secondary metabolite production is activated at specific developmental stages of *Streptomyces* life cycle. Despite this, *Streptomyces* differentiation in liquid nonsporulating cultures (flasks and industrial bioreactors) tends to be underestimated and the most important parameters managed are only indirectly related to differentiation: modifications to the culture media, optimization of productive strains by random or directed mutagenesis, analysis of biophysical parameters, etc. In this chapter, we review the relationship between differentiation and antibiotic production in liquid cultures. Morphological differentiation in liquid cultures is comparable to that occurring during presporulation stages in solid cultures: an initial compartmentalized mycelium suffers a programmed cell death, and remaining viable segments then differentiate to a second multinucleated antibiotic-producing mycelium. Differentiation is one of the keys to interpreting biophysical fermentation parameters and to rationalizing the optimization of secondary metabolite production in liquid cultures.

Keywords: *Streptomyces*, bioreactor, differentiation, antibiotics, programmed cell death

1. Introduction

Streptomycetes are gram-positive, environmental soil bacteria that play important roles in the mineralization of organic matter. *Streptomyces* is extremely important in biotechnology, given that approximately two thirds of all clinical antibiotics and several other bioactive compounds are synthesized by members of this genus [1].

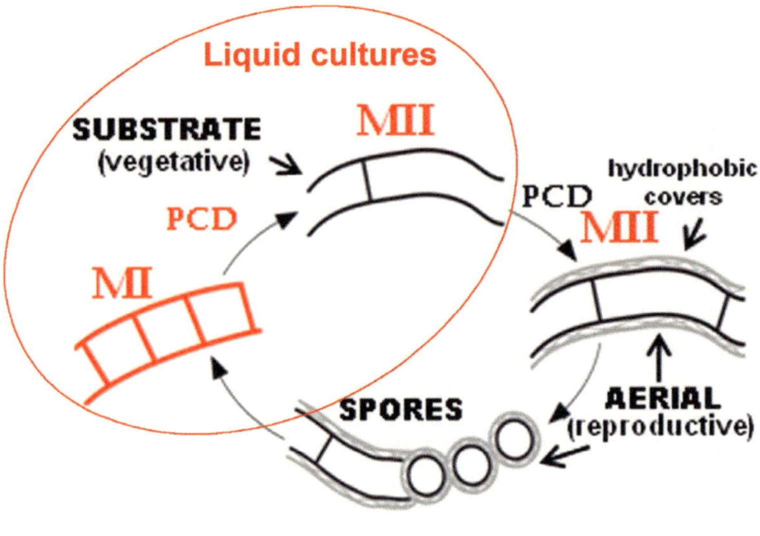

Figure 1.
Streptomyces developmental cycle. New developmental stages are highlighted in red. MI, first compartmentalized mycelium, vegetative; MII, second multinucleated mycelium producing secondary metabolites. The full cycle is developed in solid cultures. Liquid cultures do not develop aerial mycelium or spores.

Streptomycetes have a complex developmental cycle that makes this bacterium a multicellular prokaryotic model. The classical *Streptomyces* developmental cycle defined 50 years ago by Waskman [2] and Wildermuth [3] in laboratory solid cultures focused on sporulation (**Figure 1**): after spore germination, hypha grow inside the culture (substrate mycelium); the substrate mycelium differentiates into a new mycelium that begins to express hydrophobic surfaces and grows into the air (aerial mycelium); the substrate and aerial mycelia are multinucleated with sporadic septa; both mycelia eventually undergo a programmed cell death (PCD), and the remaining viable hyphae differentiate into chains of unigenomic spores. Most streptomycetes do not sporulate in liquid cultures, and it was traditionally thought that there was no differentiation in these conditions. Despite this, most processes for secondary metabolite production are performed in liquid (flask or bioreactor). It was postulated that secondary metabolites would be produced in liquid cultures by the substrate mycelium at the stationary stage [4–8].

The main objective of this work is to review the state of the art of *Streptomyces* differentiation in liquid cultures, especially in lab-scale bioreactors, defining the kind of differentiation present under these conditions, how differentiation, fermentation parameters (dissolved oxygen tension, oxygen uptake rate, oxygen transfer rate, pH, temperature, agitation, culture medium, etc.) and secondary metabolite production are correlated, and describing a general model applicable to improving secondary metabolite production in *Streptomyces* industrial fermentations.

2. *Streptomyces* development in liquid cultures

During the last decade, new knowledge regarding *Streptomyces* development during the early differentiation stages, those occurring between spore germination and substrate mycelium differentiation, was generated (red labels in **Figure 1**): a previously unidentified, young, compartmentalized mycelium (MI) undergoes a PCD; and viable MI segments differentiate into a multinucleated mycelium with sporadic septa (MII). One-micron spaced MI compartments are separated by cross-membranes without peptidoglycan cell walls, a kind of cell division unprecedented in bacteria [9]. MII corresponds to the substrate mycelium in early development, and to the aerial mycelium once it starts to express hydrophobic surfaces. In liquid cultures, there is no hydrophobic surface formation or sporulation, but there is a compartmentalized MI, which differentiates into a multinucleated MII after PCD. MII produces secondary metabolites in solid and liquid cultures. This was the first time that secondary metabolism was associated with a specific mycelial stage (MII) [10]. Transcriptomic [11] and proteomic [12, 13] analyses performed by our group corroborated that MI is the vegetative mycelium expressing/translating primary metabolism genes/proteins, while MII is the reproductive mycelium expressing/translating secondary metabolism and sporulation genes/proteins.

Microbial multinucleated structures such as the substrate and aerial mycelia are fragile, uncommon, and usually related to transitory reproductive stages [14]. Contrary to what was postulated during the last 50 years, *Streptomyces* is not an exception, and the compartmentalized MI is the predominant mycelium in cultures resembling natural conditions, such as nonamended soils [15]. The MI lifespan is very short in laboratory cultures, which is why it was ignored in most *Streptomyces* works [15].

3. *Streptomyces* differentiation and industrial fermentations

The absence of an understanding of *Streptomyces* differentiation in liquid cultures has long precluded the existence of a general consensus as to how morphological and biophysical parameters correlate with secondary metabolite production. Pharmaceutical companies have addressed the optimization of industrial fermentation empirically for each strain and compound. For example, pellet and clump formation has been described as essential for obtaining good production of retamycin or nikkomycin [16], but in the case of virginiamycin, there is no relationship between morphology and secondary metabolite production [17]; high dissolved oxygen tensions (DOT) have been reported as being necessary for the production of vancomycin [18], but not for the production of erythromycin [19], just to name a few examples.

Only recently has basic knowledge about differentiation in liquid cultures been generated. As introduced above, MII was demonstrated to be the antibiotic-producing mycelium. In a recent work, our group demonstrated that differentiation is one of the keys to interpreting typical fermentation parameters (growth, antibiotic production, dissolved oxygen tension, agitation, and oxygen uptake rates) [20] in bioreactors. Pellet and clump formation greatly influences PCD, usually occurring in the center of the mycelium pellet [10] and MII differentiation from MI living cells at the pellet periphery [10]. We proposed a general consensus to improve secondary metabolite production in *S. coelicolor*: optimization of the differentiation of the antibiotic-producing mycelium (MII) [20]. In the past few years, other groups have contributed to identifying some genetic determinants controlling pellet and clump formation [21, 22].

3.1 Differentiation of *Streptomyces* in bioreactors

There are important differences between liquid cultures in laboratory flasks and bioreactors. Despite the obvious hydrodynamic differences between the two systems, there are also differences in the culture media. In this sense, one of the most important differences between flasks and bioreactors is the use of antifoams. Antifoams are often used in bioreactors to prevent foam formation and its interference with the bioreactor probes [23]. Apparently, antifoams do not affect development and they are usually added automatically in small amounts when foam is detected by a specific probe, and in some cases, they are added directly to the culture **medium** at concentrations up to 0.1% [23]. Rioseras et al. [20] analyzed for the first time *Streptomyces* differentiation (MI and MII differentiation) in bioreactors and demonstrated an important effect of the antifoam in macroscopic morphology of the cultures (pellet and clump formation), which conditions mycelium differentiation and secondary metabolite production.

3.1.1 Differentiation of S. coelicolor in antifoam-free media

Mycelium differentiation in bioreactors is comparable to differentiation in laboratory flasks [20] (outlined in **Figure 2**): at early time points, hyphae presented the regular discontinuities and gaps previously described for MI hyphae [10]; MI differentiates into a second multinucleated mycelium (MII) after a programmed cell death (reviewed in Yagüe et al. [24]). However, there are important differences between development in flask and bioreactor.

One of the most important differences observed in the bioreactor with respect to laboratory flasks for *S. coelicolor* [20] is the existence of massive fragmentation and disintegration of mycelial pellets at around 50 hours of fermentation. This massive disintegration is observed macroscopically, in the form of the apparent clarification of the culture medium, and correlates with a sudden fall in intracellular protein levels. This kind of mycelial disintegration has been previously described as "massive lysis" in several *Streptomyces* fermentations, such as *S. clavuligerus* (reducing mycelium by more than 30%) [25], *Streptomyces* spp. [26], *Streptomyces albulus* [27], or *S. coelicolor* [28], to name just a few examples. This "massive lysis" differs from the "fragmentation of the mycelial clumps" described in some cases [29], flasks and bioreactors, which basically consists of the fragmentation of large clumps into small clumps, but without the early massive hyphal lysis reported in several bioreactor

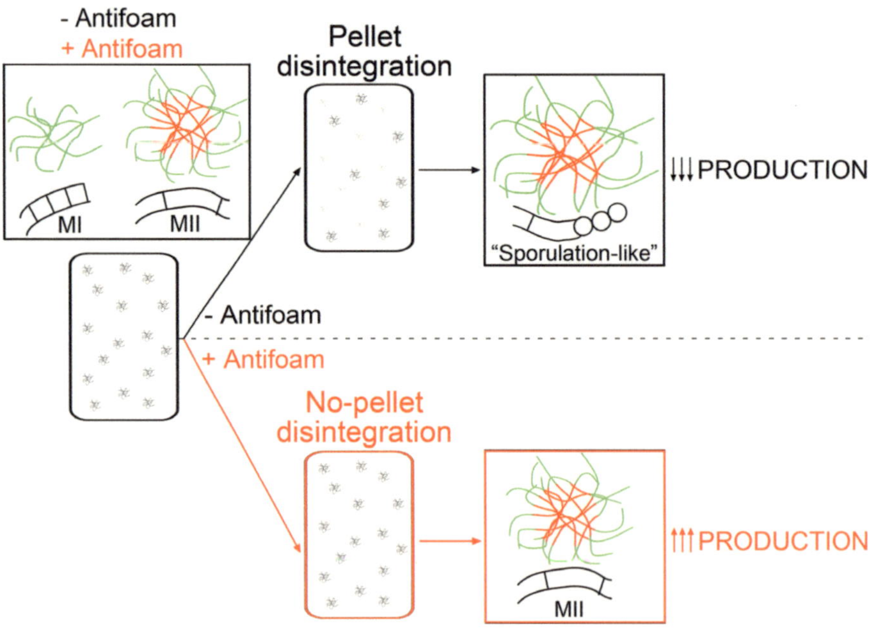

Figure 2.
Scheme illustrating S. coelicolor differentiation in bioreactors in media with or without antifoam. Red corresponds to dying hyphae (PI staining) and green to viable hyphae (SYTO9 staining). The optimal fermentation workflow is highlighted in red.

fermentations [20, 25–28]. The reason why this phenomenon occurred in some streptomycetes and not in others remains unknown. Understanding and controlling mycelial lysis would be essential to optimize fermentations. In *S. coelicolor* growing in laboratory flasks containing the same media used by Rioseras et al. [20] in bioreactors, this massive lysis does not occur [10]. Hence, Rioseras et al. [20] proposed that massive pellet disintegration depends on the hydrodynamics of the bioreactor combined with the tendency of *S. coelicolor* to form large pellets. Further work will be necessary to define if the same is happening in the other streptomycetes suffering the same phenomenon in bioreactor cultures.

MII differentiation and antibiotic production is accelerated in the *S. coelicolor* bioreactor cultures, peaking at 100–140 hours in laboratory flasks vs. 50 hours in the bioreactor [20]. However, antibiotic biosynthesis was halted after pellet disintegration, with maximum undecylprodigiosin and actinorhodin production levels lower in the bioreactor than in laboratory flasks [20].

Biophysical fermentation parameters, such as dissolved oxygen tension (DOT), agitation, and oxygen uptake rates (OURs), correlated well with differentiation [20]: DOT falls from saturation at time zero to the fixed level (50% saturation), due to hyphal growth and respiration; there is a concomitant increase in agitation to maintain oxygen levels at the fixed level; once pellet disintegration starts, biological oxygen consumption and agitation decrease gradually, and dissolved oxygen levels increase suddenly to saturation. OUR values fall during the MI PCD and are not recovered until the MII differentiation. This description of mycelium differentiation and OUR in bioreactors constitutes an elegant example of the importance of understanding *Streptomyces* differentiation, to interpret classical biophysical fermentation parameters in the model strain *S. coelicolor* and conceivably in other industrial relevant streptomycetes. Information concerning oxygen uptake kinetics of *Streptomyces* cultures is scarce despite their industrial importance. OUR values vary widely between strains, from 2.88 mg O_2 g $cell^{-1}$ h^{-1} in *S. lividans* [30] to 320 mg O_2 g $cell^{-1}$ h^{-1} in *S. clavuligerus* [31]. The meaning of these differences is difficult to interpret due to the absence of any indication as to mycelium differentiation/PCD in most of these works. An analysis of hyphae differentiation, development, and PCD would be essential to address these differences in OURs between different *Streptomyces* strains and culture conditions.

3.1.2 Differentiation of S. coelicolor in bioreactors supplemented with antifoam

Rioseras et al. [20] modified growing conditions in bioreactors to prevent the early massive lysis described above. The most obvious difference between bioreactors and laboratory flasks is the impellers used for agitation in the case of the bioreactor, so the first experimental approach to trying to prevent lysis was to reduce agitation to minimum levels (50 rpm); however, the same extension of pellet disintegration was observed [20]. Similar results were observed at different agitation rates (50, 100, 200, or 300 rpm) or by replacing Rushton impellers by a gentle impeller (pitched blade impellers) [20]. The only modification that worked to avoid the massive mycelial lysis observed in bioreactors was the modification of the culture medium's rheology reducing surface tension by means of an antifoam agent (Biospumex 153 K, BASF). This effect of preventing early fragmentation/lysis was only observed at relatively high antifoam concentrations (1%) (outlined in **Figure 2**).

The reason why antifoam prevents pellet disintegration is as yet unknown. However, the antifoam tended to coat the mycelial pellets, and the hydrophobic forces generated may have prevented this phenomenon. Antifoams are often used with *Streptomyces coelicolor* [23] as well as other *Streptomyces* fermentations to prevent foam formation, or even, in some cases to be used as carbon sources [32].

They are usually added automatically in small amounts when foam is detected by a specific probe, and in some cases, they are added directly to the culture medium at concentrations up to 0.1% [23]. However, Rioseras et al. [20] were the first authors who demonstrated that antifoams added to the culture media at relatively high concentrations prevent mycelium lysis, a fact that might be useful for preventing lysis in other industrial streptomycetes.

Biophysical fermentation parameters also correlated well with differentiation in cultures with antifoam: the absence of pellet disintegration prolonged the oxygen consumption phase, generating two peaks of OUR (MI and MII stages) separated by a stage of low oxygen consumption [20]. These two maxima in OUR are very unusual in industrial fermentation, and are another nice illustration of the necessity of understanding *Streptomyces* differentiation in order to interpret fermentation parameters in *Streptomyces* fermentations. As in the case of fermentations without antifoam, oxygen levels did not limit growth.

3.2 Sporulation of *Streptomyces* in bioreactors

Another important difference between bioreactor- and laboratory flask- S. *coelicolor* cultures is the existence of a sporulation-like process, affecting some 5% of hyphae in bioreactors [20]. Two of the most important features of sporulation, division, and separation of nucleoids, and the physical strangulation of hypha forming chains of individual round segments, were observed in bioreactors [20]. Sporulation in *S. coelicolor* liquid cultures is very unusual and has only been reported once before in laboratory flasks suffering nutritional downshifts [33]. The differentiation signals activating sporulation in the bioreactors remain unknown. However, if it is considered that sporulation is triggered by environmental/biological stresses [34], the high growth rates achieved in the bioreactors together with pellet disintegration might approach the development occurring in stressed solid sporulating cultures. In the absence of pellet disintegration, putative differentiation diffusible signals [35, 36] generated by stressed cells suffering from PCD [24] would be hidden in the centers of the pellets [20].

4. *Streptomyces* differentiation and screening for new bioactive compounds

Streptomycetes are important biotechnological bacteria from which two thirds of the bioactive secondary metabolites used clinically (mainly antibiotics, but also antitumourals, immunosuppressors, etc.) were discovered [37]. Drug discovery became challenging once the most common antibiotics were discovered. In fact, during the past 30 years, only three new classes of antibiotics have been brought to the clinic (mutilins, lipopeptides, and oxazolidinones) [38, 39]. At the same time, microbial resistance to existing antibiotics has increased dramatically, rendering some microbial infections extremely hard to treat.

New antibiotics are urgently needed in the clinic. No valid alternatives to screening natural strains have emerged to find new scaffolds and families of antibiotics [40]. New workflows are needed to access the natural secondary metabolites that remain inaccessible in the laboratory [41]. Nonnatural synthetic antibiotics obtained by chemical/combinatorial biosynthesis exist, but most of them are variations of natural molecules [40]. The best way to find structurally novel bioactive compounds is to resume screening from natural streptomycetes. The most obvious approach to look for new bioactive compounds from natural streptomycetes is to study streptomycetes isolated from relatively lowly explored niches such as marine ecosystems [42], symbiotic streptomycetes [43], etc., an approach followed by several research groups and biotechnology companies. On the other hand, genomic analyses revealed that Streptomyces genomes encode an average of 30 secondary metabolite pathways [44], but only a fraction of these pathways (around four per strain [43]) is active in laboratory cultures. Consequently, there is a huge amount of potentially bioactive compounds produced by streptomycetes that are never observed in the lab (cryptic pathways) and remain unexplored. There is a consensus in the scientific community about the necessity to activate the expression of these cryptic pathways in order to overcome the present bottleneck in drug discovery.

Several research groups and biotechnology companies face the challenge of activating cryptic pathways to try to mimic the ecological niche of the bacteria by making co-cultures of different microbes [45], looking for elicitor activating pathways (nutrients such as glucose, xylose, and small molecules such as GlcNac and phosphate) [46] or making heterologous expression [47]. As stated above, *Streptomyces* differentiation conditions secondary metabolism [20], and differentiation can be one of the keys to activate cryptic pathways by modulating the differentiation of the antibiotic producer mycelium (MII). The "MII approach" will be useful to activate secondary metabolism production in strains discarded as producers in the classical screening campaigns, strains that did not produce in the lab because they did not reach the MII stage, or because they sporulated in liquid cultures (during sporulation metabolism is stopped), i.e. false negatives [48]. This approach was already useful to enhance flavonoid production in *S. albus* [49], as well as microbial transglutaminase in *S. mobaraensis* [50].

5. Conclusions

Different streptomycetes show different behaviors in liquid cultures: some species form large pellets, such as *S. coelicolor*, others grow more dispersed, as for instance *S. clavuligerus* [25], while some species such as *S. griseus* or *S. venezuelae* sporulate in liquid cultures [51, 52]. As a consequence, the effect of fermentation parameter modifications in different species cannot be easily predicted. The "MII approach" described here, i.e. optimization of antibiotic-producing mycelium differentiation, prevention of sporulation, might be applied to rationalizing the biological effects of classical biophysical fermentation parameters, and to facilitating the optimization of secondary metabolite production in industrial streptomycetes. In addition, preventing early massive pellet fragmentation/lysis by adding antifoam directly to the culture medium at relatively high concentrations is novel and may be useful for preventing lysis in other industrial streptomycetes. The "MII approach" also has applications in the screening for new secondary metabolites, allowing the activation of cryptic pathways in streptomycetes that did not reach the MII stage in the lab.

Acknowledgements

We thank the Spanish "Ministerio de Economía y Competitividad" (MINECO; BIO2015-65709-R) and the "Marie Curie cofund Clarin" Grant for financial support and Proof-Reading-Service.com for proofreading the final manuscript.

Conflict of interest

There is no conflict of interest in this work.

Author details

Angel Manteca*, Beatriz Rioseras, Nathaly González-Quiñónez,
Gemma Fernández-García and Paula Yagüe
Área de Microbiología, Departamento de Biología Funcional e IUOPA, Facultad de Medicina, Universidad de Oviedo, Oviedo, Spain

*Address all correspondence to: mantecaangel@uniovi.es

References

[1] Ruiz B, Chavez A, Forero A, Garcia-Huante Y, Romero A, Sanchez M, et al. Production of microbial secondary metabolites: Regulation by the carbon source. Critical Reviews in Microbiology. 2010;**36**(2):146-167

[2] Waksman SA. The Actinomycetes: Their Nature, Occurence, Activities and Importance. Waltham, Mass: Chronica Botanica Co; 1950

[3] Wildermuth H. Development and organization of the aerial mycelium in *Streptomyces coelicolor*. Journal of General Microbiology. 1970;**60**(1):43-50

[4] Chouayekh H, Nothaft H, Delaunay S, Linder M, Payrastre B, Seghezzi N, et al. Phosphoinositides are involved in control of the glucose-dependent growth resumption that follows the transition phase in *Streptomyces lividans*. Journal of Bacteriology. 2007;**189**(3):741-749

[5] Granozzi C, Billetta R, Passantino R, Sollazzo M, Puglia AM. A breakdown in macromolecular synthesis preceding differentiation in *Streptomyces coelicolor* A3(2). Journal of General Microbiology. 1990;**136**(4):713-716

[6] Novotna J, Vohradsky J, Berndt P, Gramajo H, Langen H, Li XM, et al. Proteomic studies of diauxic lag in the differentiating prokaryote *Streptomyces coelicolor* reveal a regulatory network of stress-induced proteins and central metabolic enzymes. Molecular Microbiology. 2003;**48**(5):1289-1303

[7] Vohradsky J, Li XM, Thompson CJ. Identification of procaryotic developmental stages by statistical analyses of two-dimensional gel patterns. Electrophoresis. 1997;**18**(8):1418-1428

[8] Zhou LH, Li YQ, Li YQ, Wu D. Spatio-temporal expression of the pathway-specific regulatory gene redD in *S. coelicolor*. Journal of Zhejiang University. Science. 2005;**B** (6, 6):464-469

[9] Yague P, Willemse J, Koning RI, Rioseras B, Lopez-Garcia MT, Gonzalez-Quinonez N, et al. Subcompartmentalization by cross-membranes during early growth of *Streptomyces hyphae*. Nature Communications. 2016;7:12467

[10] Manteca A, Alvarez R, Salazar N, Yague P, Sanchez J. Mycelium differentiation and antibiotic production in submerged cultures of *Streptomyces coelicolor*. Applied and Environmental Microbiology. 2008;**74**(12):3877-3886

[11] Yague P, Rodriguez-Garcia A, Lopez-Garcia MT, Rioseras B, Martin JF, Sanchez J, et al. Transcriptomic analysis of liquid non-sporulating *Streptomyces coelicolor* cultures demonstrates the existence of a complex differentiation comparable to that occurring in solid sporulating cultures. PLoS One. 2014;**9**(1):e86296

[12] Manteca A, Sanchez J, Jung HR, Schwammle V, Jensen ON. Quantitative proteomics analysis of *Streptomyces coelicolor* development demonstrates that onset of secondary metabolism coincides with hypha differentiation. Molecular & Cellular Proteomics. 2010;**9**(7):1423-1436

[13] Rioseras B, Shliaha PV, Gorshkov V, Yague P, Lopez-Garcia MT, Gonzalez-Quinonez N, et al. Quantitative proteome and phosphoproteome analyses of *Streptomyces coelicolor* reveal proteins and phosphoproteins modulating differentiation and secondary metabolism. Molecular & Cellular Proteomics. 2018;**17**(8):1591-1611

[14] Daniels R, Vanderleyden J, Michiels J. Quorum sensing and

swarming migration in bacteria. FEMS Microbiology Reviews. 2004;**28**(3):261-289

[15] Manteca A, Sanchez J. Streptomyces development in colonies and soils. Applied and Environmental Microbiology. 2009;**75**(9):2920-2924

[16] Pamboukian CR, Facciotti MC. Production of antitumoral retamycin during fed-batch fermentations of *Streptomyces olindensis*. Applied Biochemistry and Biotechnology. 2004;**112**(2):111-122

[17] Yang YK, Morikawa M, Shimizu H, Shioya S, Suga KI, Nihira T, et al. Image analysis of mycelial morphology in virginiamycin production by batch culture of *Streptomyces virginiae*. Journal of Fermentation and Bioengineering. 1996;**81**:7-12

[18] Dunstan GH, Avignone-Rossa C, Langley D, Bushell ME. The vancomycin biosynthetic pathway is induced in oxygen-limited *Amycolatopsis orientalis* (ATCC 19795) cultures that do not produce antibiotic. Enzyme and Microbial Technology. 2000;**27**:502-510

[19] Clark GJ, Langley D, Bushell ME. Oxygen limitation can induce microbial secondary metabolite formation, investigations with miniature electrodes in shaker and bioreactor culture. Microbiology. 1995;**141**:663-669

[20] Rioseras B, Lopez-Garcia MT, Yague P, Sanchez J, Manteca A. Mycelium differentiation and development of *Streptomyces coelicolor* in lab-scale bioreactors: Programmed cell death, differentiation, and lysis are closely linked to undecylprodigiosin and actinorhodin production. Bioresource Technology. 2014;**151**:191-198

[21] Petrus ML, Vijgenboom E, Chaplin AK, Worrall JA, van Wezel GP, Claessen D. The DyP-type peroxidase DtpA is a Tat-substrate required for GlxA maturation and morphogenesis in Streptomyces. Open Biology. 2016;**6**(1):150149

[22] Zacchetti B, Willemse J, Recter B, van Dissel D, van Wezel GP, Wosten HA, et al. Aggregation of germlings is a major contributing factor towards mycelial heterogeneity of Streptomyces. Scientific Reports. 2016;**6**:27045

[23] Wentzel A, Bruheim P, Overby A, Jakobsen OM, Sletta H, Omara WA, et al. Optimized submerged batch fermentation strategy for systems scale studies of metabolic switching in *Streptomyces coelicolor* A3(2). BMC Systems Biology. 2012;**6**:59

[24] Yague P, Lopez-Garcia MT, Rioseras B, Sanchez J, Manteca A. Pre-sporulation stages of Streptomyces differentiation: State-of-the-art and future perspectives. FEMS Microbiology Letters. 2013;**342**(2):79-88

[25] Roubos JA, Krabben P, Luiten RG, Verbruggen HB, Heijnen JJ. A quantitative approach to characterizing cell lysis caused by mechanical agitation of *Streptomyces clavuligerus*. Biotechnology Progress. 2001;**17**(2):336-347

[26] Techapun C, Poosaran N, Watanabe M, Sasaki K. Optimization of aeration and agitation rates to improve cellulase-free xylanase production by thermotolerant *Streptomyces* sp. Ab106 and repeated fed-batch cultivation using agricultural waste. Journal of Bioscience and Bioengineering. 2003;**95**(3):298-301

[27] Shih IL, Shen MH. Optimization of cell growth and poly(ε-lysine) production in batch and fed-batch cultures by *Streptomyces albulus* IFO 14147. Process Biochemistry. 2006;**41**:1644-1649

[28] Ozergin-Ulgen K, Mavituna F. Actinorhodin production by

Streptomyces coelicolor A3(2): Kinetic parameters related to growth, substrate uptake and production. Applied Microbiology and Biotechnology. 1993;**40**:457-462

[29] van Wezel GP, Krabben P, Traag BA, Keijser BJ, Kerste R, Vijgenboom E, et al. Unlocking *Streptomyces* spp. for use as sustainable industrial production platforms by morphological engineering. Applied and Environmental Microbiology. 2006;**72**(8):5283-5288

[30] Magnolo SK, Leenutaphong DL, DeModena JA, Curtis JE, Bailey JE, Galazzo JL, et al. Actinorhodin production by *Streptomyces coelicolor* and growth of *Streptomyces lividans* are improved by the expression of a bacterial hemoglobin. Biotechnology (N. Y). 1991;**9**(5):473-476

[31] Yegneswaran PK, Gray MR, Thompson BG. Experimental simulation of dissolved oxygen fluctuations in large fermentors: Effect on *Streptomyces clavuligerus*. Biotechnology and Bioengineering. 1991;**38**(10):1203-1209

[32] Perlman D, Wagman GH. Studies on the utilization of lipids by *Streptomyces griseus*. Journal of Bacteriology. 1952;**63**(2):253-262

[33] Daza A, Martin JF, Dominguez A, Gil JA. Sporulation of several species of Streptomyces in submerged cultures after nutritional downshift. Journal of General Microbiology. 1989;**135**:2483-2491

[34] Chater KF. Regulation of sporulation in *Streptomyces coelicolor* A3(2): A checkpoint multiplex? Current Opinion in Microbiology. 2001;**4**(6):667-673

[35] Chater KF, Biro S, Lee KJ, Palmer T, Schrempf H. The complex extracellular biology of Streptomyces. FEMS Microbiology Reviews. 2010;**34**(2):171-198

[36] Horinouchi S, Beppu T. Autoregulatory factors and communication in actinomycetes. Annual Review of Microbiology. 1992;**46**:377-398

[37] Berdy J. Bioactive microbial metabolites. Journal of Antibiotics (Tokyo). 2005;**58**(1):1-26

[38] Fischbach MA, Walsh CT. Antibiotics for emerging pathogens. Science. 2009;**325**(5944):1089-1093

[39] Hopwood DA. Streptomyces in Nature and Medicine: The Antibiotic Makers. New York, Oxford: Oxford University Press; 2007

[40] Weissman KJ, Leadlay PF. Combinatorial biosynthesis of reduced polyketides. Nature Reviews. Microbiology. 2005;**3**(12):925-936

[41] Onaka H. Novel antibiotic screening methods to awaken silent or cryptic secondary metabolic pathways in actinomycetes. Journal of Antibiotics (Tokyo). 2017;**70**(8):865-870

[42] Manivasagan P, Kang KH, Sivakumar K, Li-Chan EC, Oh HM, Kim SK. Marine actinobacteria: An important source of bioactive natural products. Environmental Toxicology and Pharmacology. 2014;**38**(1):172-188

[43] Parrot D, Legrave N, Delmail D, Grube M, Suzuki M, Tomasi S. Review—Lichen-associated bacteria as a hot spot of chemodiversity: Focus on uncialamycin, a promising compound for future medicinal applications. Planta Medica. 2016;**82**(13):1143-1152

[44] Genilloud O. The re-emerging role of microbial natural products in antibiotic discovery. Antonie Van Leeuwenhoek. 2014;**106**(1):173-188

[45] Marmann A, Aly AH, Lin W, Wang B, Proksch P. Co-cultivation—A powerful emerging tool for

enhancing the chemical diversity of microorganisms. Marine Drugs. 2014;**12**(2):1043-1065

[46] Liu G, Chater KF, Chandra G, Niu G, Tan H. Molecular regulation of antibiotic biosynthesis in streptomyces. Microbiology and Molecular Biology Reviews. 2013;**77**(1):112-143

[47] Gomez-Escribano JP, Bibb MJ. Engineering *Streptomyces coelicolor* for heterologous expression of secondary metabolite gene clusters. Microbial Biotechnology. 2011;**4**(2):207-215

[48] Manteca A, Yague P. Streptomyces differentiation in liquid cultures as a trigger of secondary metabolism. Antibiotics (Basel). 2018;**7**(2)

[49] Marin L, Gutierrez-Del-Rio I, Yague P, Manteca A, Villar CJ, Lombo F. De novo biosynthesis of apigenin, luteolin, and eriodictyol in the actinomycete *Streptomyces albus* and production improvement by feeding and spore conditioning. Frontiers in Microbiology. 2017;**8**:921

[50] Treppiccione L, Ottombrino A, Luongo D, Maurano F, Manteca A, Lombó F, et al. Development of gluten with immunomodulatory properties using mTG-active food grade supernatants from *Streptomyces mobaraensis* cultures. Journal of Functional Foods. 2017;**34**:390-397

[51] Kendrick KE, Ensign JC. Sporulation of *Streptomyces griseus* in submerged culture. Journal of Bacteriology. 1983;**155**(1):357-366

[52] Glazebrook MA, Doull JL, Stuttard C, Vining LC. Sporulation of *Streptomyces venezuelae* in submerged cultures. Journal of General Microbiology. 1990;**136**(3):581-588